Jewish Resistance during the Holocaust

Also by James M. Glass

DELUSION

'LIFE UNWORTHY OF LIFE': Racial Phobia and Mass Murder in Hitler's Germany

PRIVATE TERROR/PUBLIC LIFE

PSYCHOSIS AND POWER

SHATTERED SELVES

Jewish Resistance during the Holocaust
Moral Uses of Violence and Will

James M. Glass
Department of Government and Politics
University of Maryland, College Park

© James M. Glass 2004
Softcover reprint of the hardcover 1st edition 2004 978-1-4039-3907-4
All rights reserved. No reproduction, copy or transmission of this
publication may be made without written permission.

No paragraph of this publication may be reproduced, copied or transmitted
save with written permission or in accordance with the provisions of the
Copyright, Designs and Patents Act 1988, or under the terms of any licence
permitting limited copying issued by the Copyright Licensing Agency, 90
Tottenham Court Road, London W1T 4LP.

Any person who does any unauthorized act in relation to this publication
may be liable to criminal prosecution and civil claims for damages.

The author has asserted his right to be identified as the author of this
work in accordance with the Copyright, Designs and Patents Act 1988.

First published 2004 by
PALGRAVE MACMILLAN
Houndmills, Basingstoke, Hampshire RG21 6XS and
175 Fifth Avenue, New York, N.Y. 10010
Companies and representatives throughout the world

PALGRAVE MACMILLAN is the global academic imprint of the Palgrave
Macmillan division of St. Martin's Press, LLC and of Palgrave Macmillan Ltd.
Macmillan® is a registered trademark in the United States, United Kingdom
and other countries. Palgrave is a registered trademark in the European
Union and other countries.

ISBN 978-1-349-51947-7   ISBN 978-0-230-50013-6 (eBook)

DOI 10.1057/9780230500136

This book is printed on paper suitable for recycling and made from fully
managed and sustained forest sources.

A catalogue record for this book is available from the British Library.

Library of Congress Cataloging-in-Publication Data
Glass, James M.
   Jewish resistance during the Holocaust: moral uses of violence and will/
James M. Glass.
      p. cm.
   Includes bibliographical references and index.

   1. World War, 1939–1945–Jewish resistance–Moral and ethical
aspects. 2. Violence–Moral and ethical aspects. 3. Holocaust, Jewish
(1939–1945)–Moral and ethical aspects. I. Title.
D810.J4.G565 2004
940.53′1832–dc22   2004044372

10 9 8 7 6 5 4 3 2 1
13 12 11 10 09 08 07 06 05 04

For the resistance survivors and in remembrance of their families and comrades, and for the memory of Rabbi Kalonymus Kalman Shapira of Warsaw who fought with spirit and will.

Contents

Acknowledgements		ix
Introduction: Memory, Resistance and Reclaiming the Self		1
1	The Moral Justification for Killing	9
2	Collective Trauma: The Disintegration of Ethics	27
3	The Moral Position of Violence: Bielski Survivors	55
4	The Moral Goodness of Violence: Necessity in the Forests	79
5	Spiritual Resistance: Understanding its Meaning	103
6	Condemned Spirit and the Moral Arguments of Faith	121
7	The Silence of Faith Facing the Emptied-out Self	141
8	Law and Spirit in Terrible Times	155
Notes		169
Bibliography		189
Index		199

Acknowledgements

I am grateful to a number of individuals and organizations for invaluable aid and assistance during the writing of this book. I would like to thank the Simon Wiesenthal Center and the Shoah Foundation, both in Los Angeles, for their help in locating resistance survivors. I am also indebted to Wendy Lower and Paul Shapiro and their staff, of the Center for Advanced Holocaust Studies at the United States Holocaust Memorial Museum in Washington, D.C., for allowing me access to the library, archives and working space within the Museum. Also I would like to thank Michael Haley-Goldman for his archival assistance at the Museum.

Pieces of this research were developed in a number of invited presentations: The George Washington University Seminar in Political Psychology and Leadership (The *Judenrat* and Collaboration), December 1997; The Washington School of Psychiatry, Seminar in Group Psychology (Resistance, Collaboration and Groups in the Holocaust), January 1998; The Sheppard and Enoch Pratt Hospital: Grand Rounds Lecture (The Impact of Violence on the Self), July 1998; The United States Holocaust Conference in Millersville, PA, Plenary Session (Mass Murder and the Actions of Jewish Resistance), April 1998; The Austin Riggs Center, Conference on 'Psychosis and its Social Context', Stockbridge, MA (Resistance, Madness and Voices of Sanity), October 1998; the Conference for the Study of Organizational Change, University of Missouri, Columbia, MO (The Psychology of Organization in Jewish Resistance during the Holocaust), September 1998; the United States Holocaust Memorial Museum, Center for Advanced Holocaust Studies. Noon Lecture Series (Franz Fanon's Theory of Violence and Jewish Resistance during the Holocaust), May 1999; The United States Department of State, Secretary's Open Forum (The Psychology of Genocide and the Psychology of Resistance), June 1999; the United States Department of State, Foreign Service Institute, Alexandria, VA (The Concept of 'Worthless Life' and the Jewish Resistors' Reaction and Action), October 2000; California State University at Long Beach, Lecture Series (Moral Issues in Jewish Resistance and Annihilatory Violence), November 2002.

Mitch Branff, of the Jewish Partisan Educational Foundation in San Francisco, was most generous in helping me locate resistance survivors and in speaking to me about their experiences. Thanks are also due to Shaare Tefile congregation in Silver Spring, Maryland and B'nai Shalom congregation in Olney, Maryland for the opportunity to speak with the members about Jewish resistance and the psychological position of the survivor. For research support and funding I would like to thank the Provost's Office of the University of Maryland College Park for the Distinguished Scholar Teacher Award, 2002–3, and its research stipend.

I am indebted to conversations with Miles Lerman, founder of the Miles Lerman Center for the Study of Jewish Resistance at the United States Holocaust Memorial Museum, and Yitzhak Arad, both resistance survivors, who gave me valuable insights into the facts and experience of resistance and its psychology. Also, the research would not have been possible without the cooperation and conversation with the many resistance survivors who graciously offered me their time, hospitality and recollection. I am particularly grateful to Shlomo Berger, Vernon Rusheen, Ben Kamm, Sonia Bielski, Sonia Oshman, Leah Johnson, Charles and Sarah Bedzow, Aaron (Bielski) Bell, Elsi Shor, Frank Blaichman and Simon Trakinski. These resistance survivors were not only generous with their time, but willing to respond to questions that moved into emotionally troubling areas. I would also like to thank Zvi Bielski who gave me invaluable insights into the actions of his father, Zush Bielski, and his uncle, Tuvia Bielski.

A number of friends and colleagues commented on different phases of the typescript and its development. I would like to thank Bob Alperin, Benjamin Barber, Mary Caputi, Michael Diamond, Jane Flax, Roger Haydon, Jerrold Post, Seymour Rubenfeld, Mark Warren and Victor Wolfenstein. Mark Lichbach and Fred Alford read the entire typescript, and their suggestions were critical in the various permutations of the approach and the ethical, religious and theological issues pertaining to Jewish resistance. I would also like to thank my typist Flora Paoli who, over the years, has been an inspiring source of information, technical intelligence and editing. Her painstaking care in transforming scribbled-over text into a neat typescript has been an art that guides and informs my writing.

Finally, without the patience and tolerance of my family – my wife, Cyndi, and my sons, Jeremy and Jason – I never would have had the emotional space to complete this study. For their understanding and occasional interruptions, which unknown to them were essential for my own mental equilibrium over the course of writing this book, I am grateful.

Introduction: Memory, Resistance and Reclaiming the Self

I speak with resistance survivors, in their late seventies and early eighties, in New York, Los Angeles, Miami, Philadelphia, listening to their extraordinary stories and the trajectories of their lives. When they bring back the past, the setting changes and the words take the 'Then' into the 'Now.' Their stories move through the reach of the past and the pull of the present without boundary; their narratives possess the sense of literally being there.

They show me snapshots, fading black-and-white images of treasured memories, pictures of young men in uniform, groups of soldiers, men and women barely out of their teens, holding rifles; two friends smiling, their arms round each other's shoulders; the lovely girlfriend looking with affection towards the camera, leaning against a tree; a set of train tracks blown to pieces, and the commentary behind the pictures. Ben: 'You know that girl. Was she pretty; she liked me and we were very close. But after the fighting stopped, she disappeared; I think she went back to Russia. I never heard from her; I don't even think she knew my real name.' The memories speak of smuggling people out of and goods into the ghetto, enduring imprisonment in stinking jails; witnessing friends and family shot and beaten; and in turn being beaten and threatened; evading Germans; trying to deal with local populations; narrow escapes; finding homes destroyed and parents missing; killing Germans and collaborators. As these stories evolve, the age in their faces disappears and the voice of the fighter emerges, a smoldering rage, then dejection as the recollection of despair and loss consumes their consciousness. I see enormous pride and dignity and an uncompromising attitude towards the Germans and their Polish,

Ukrainian and Lithuanian sympathizers. These ex-partisans speak of having killed without guilt; what they most regret is not having killed more Germans. And always the remembrance of despair and outrage. 'What do you mean a God? How can you believe in God after you've seen a German soldier split a baby in two with rifle fire?'

These stories often rattle my composure. I stop taking notes and sit there, stone cold and terrified, watching these partisan survivors going back in time, bringing into the present images of brutality, survival, atrocity and revenge. These ex-fighters and those who had been with them, tough, now old, relive their witnessing. Why be so intimate with me, a stranger? They want me to know the scene, the struggle, the impossibility of their circumstances. Not just the facts, but more – the despair. The deeper we move into the interview, the more it seems they want me to understand what it felt like to be there. There's no apology for their rage. It needed to be affirmed, and they would not let go until I acknowledged its justice and the absolute conviction that what they did was right. It's not that they asked this of me or put it quite this way; but I could see it in their eyes and the way their bodies moved through these narratives. My own reaction to their actions had nothing to do with politeness or emotional coercion. The facts were so brutal, how could anyone listening to this past not be drawn to a conviction of the justness of their actions, of the righteousness of revenge? I felt reverence and inadequacy in their presence, but always refracting the silent question to myself: would I have been as courageous? These men and women brought to their eighties two lives: one truncated by an unimaginable violence; the other a series of struggles and blessings embodied in a discourse of sheer pride over the accomplishments of their lives in America.

How proud they were of being Jewish, not in any theological sense, but having saved an identity and, in some instances, having rescued other Jews from certain death. It was a pride in having passed on an identity without being crushed in the process. It is not the idea of being the 'chosen people'; that has no meaning for them, in the sense of possessing a special place or mandate from God. It seemed to be more like having endured and survived an outrage and preserved a history, a set of remembrances for another generation, thereby assuring the survival of that identity and its sacred words, artifacts and texts into the future. That is what they as survivors are so proud of – having acted as instruments of the Torah's enduring strength through time.

Each attributed survival, in part, to luck; and indeed luck was very much an aspect of their story. But luck figures as only a piece of the story; there is also their rage, tears, comradeship with others. The durability of resistance communities too requires acknowledgement. For these resistors, as their stories remind us time and again, community never disappeared; even if it was fighting alongside Russians or sharing meals with anti-Semites, community saved them – community as witnessing, as a band enacting revenge, as a unit undertaking combat missions. No one, they insisted, could have survived alone. Even in Auschwitz, the underground community of the resistance saved many, not each fighting for himself, which the survivors maintain defined 'ethics' in the camps. But there was more to it, more than the focus on one's own survival. In Auschwitz the underground community saved those it chose to save or believed it had to save. It ceased to matter to the Jews in the forests and the undergrounds whether they might be killed. None believed they would survive the war. Each they told me, over and over again, had moved beyond physical survival as an end in itself. What mattered more lay in retribution, killing and, for some, rescue, satisfaction in knowing they had sent the enemy to his grave or had frustrated his ability to fight, whether by sabotaging train tracks, arms depots or fuses manufactured at Auschwitz-Buna.

No guilt is expressed at having killed sympathizers; no guilt about taking whatever food they needed. These men and women became the surviving remnant for whom revenge meant saving identity and doing whatever it took to resist the oppressor's efforts to take it away. Schlomo Berger lost his outward identity as a Jew for three years; in order to fight the Germans he joined a Polish resistance group and passed himself as a Catholic, learning the religious practice and theology from a priest while imprisoned for six months by the Germans. 'I couldn't be a Jew; they would have killed me.' But the German assault on spirit never penetrated to the inner core of a self that knew who he was, what he called his 'steel fence' of faith which protected his identity and will. It was a resolve still visible in his eyes, as if he were speaking directly in the present; it is there, in his living room now and in those Polish forests that are still very much in the forefront of his consciousness.

'Then' and 'now' lose boundaries; the resistance survivors are back there, and in these interviews they want me to see what they are seeing. 'Then' breaks into the presence of the space we occupy; the violence of their memories, the unsatisfied desire for revenge, the hatred of those who collaborated and killed Jews, give their words a heavy weight that belies the tranquility of where we sit. Now, they have their children; and they want their children to see, to understand. For years, they refused to return to those anguished places in their memories; now, their pride lies in their being thankful to watch their children grow, succeed, marry, establish families, to become grandparents and great-grandparents. This is a profound delight, and in their voices is taken as a sign of victory, as if they were triumphant, saying to the Germans across the divide of time: 'You could take away so much; my childhood, my parents, my family; but here is *my* family, my world, and look at the wonder of it; look at how it has blossomed; that is certainly enough. You did not succeed, you did not win. We won, because look at what we have accomplished!' These survivors from Hell gave their children a religious education; and while for so many years they remained silent about their experience, their silence never demonstrated a loss of spirit, but was rather a survival of spirit. In the *Bar* and *Bat Mitzvah*, the Hebrew lessons and schools, what was being celebrated for these survivors was not only the triumph of their children and the mastery of difficult texts and rituals, but a cultural identity in time that could not be crushed, a special pleading for words that go back centuries and form a living presence in the rituals of Jewish rites of passage and observance. Lighting candles in Los Angeles or New York or New Jersey affirmed the practice and meaning of that history and its survival in the children of those who fought for just these very moments.

The pride in their children's accomplishments was a pride denied to their own parents; deprived of the pride of their own parents' joy, these survivors replicate that pride in what they feel towards their own children's accomplishments. It is a tribute to their lost parents, and the lost moments and times of children thrown into a universe where every imaginable security had been blasted into nothingness. Not to have a grave to mark where your parents lie; not to know how or when they died; not being able to say goodbye – none of us today knows that horror or that emptiness, the absence of being.

None of us can imagine what it is like to live with those memories, those uncertainties, and the abject feeling of not having been able to save your family. Nothing replaces that kind of severed and truncated experience of loss. Being there for their own children; having the pleasure of knowing they can witness what their parents never witnessed, to listen to the sheer joy of Ben as he describes to me his grandson's rock concert in Spain – reflections like these, which we take so much for granted, become symbols of a spiritual transcendence and a *Kaddish*, or prayer for the dead, for the survivors' own parents and lost families.

These survivor-fighters, who killed without remorse, arrived in New York or Los Angeles or Montreal with five dollars in their pocket, and, using the rage that sustained them during their resistance, transformed that energy into building new lives for themselves. As I sat in these well-ordered living rooms, houses of immaculate cleanliness, I saw not only persons who had attained every measure of worldly success, but also proud parents and despairing memories, both simultaneously expressed, but each filled with pride at having preserved their Jewish identity, and pride in knowing this identity had taken root in the next generation. I saw a respect for the country that gave them a new life, a memory for the loved ones not returning, and a rock-solid commitment to give their children a security denied their own youth. I heard then not age, but passions, traveling across time; even with the force of memory and the outrage as fresh as it was in 1945, these ex-partisans revealed an indestructible humanity and dignity.

Both the resistance fighters who did survive and the spiritual fighters who did not, such as Rabbi Kalonymus Kalman Shapira of Warsaw, acted in the world of faith, with great willfulness. There were many examples of partisans and spiritual leaders who refused to be destroyed by barbarism. Each fought in special venues to preserve a religious and cultural identity; each refused to accept the oppressor's concept of how they should be and act. Each claimed sovereignty over a psychological space they fought ferociously to protect. What is clear is that they refused the German objective of breaking down and annihilating the spirit. It is then not enough to admire the violent resistors or to acknowledge resistance only from those who fought in the forests and the underground. In the company of men and women like Miles Lerman, Sonia Bielski,

Vernon Rasheen, Sonya Oshmam, Frank Blaichman, Schlomo Berger and Ben Kamm stand spiritual leaders like Rabbi Shapira and Rabbi Oshry of Vilna, who fought for the very core of the Jewish faith: the words of the Torah and the presence of God. Even in the face of his tremendous loss, the death of his family outside a Warsaw hospital, Rabbi Shapira's sermons bear living testimony to the endurance of faith, to the capacity of the spirit to withstand oppression. It is faith that emerges victorious in these stories of violent and spiritual resistance, men and women refusing to relinquish faith in surviving horror.

It would be a mistake to dismiss the faith of men like Rabbi Shapira, to see it as a futile gesture at retaining a theological position, to relegate that voice to the silence of time. Rabbi Shapira's voice resonates through time. It was resilient and strong in Warsaw, and even though he never counseled violent resistance, he refused to bow before the German authorities. While not using the language of political power, Rabbi Shapira refused to be silenced, and his story in its own way is as compelling as those of the partisan and underground fighters. In his refusal to stop writing, to stop arguing with God, his refusal to stop delivering sermons, Rabbi Shapira, like the violent resistors, affirmed a community's place in time and history, while simultaneously denying German power over his mind and ethics. He writes in a book of instruction for his Yeshiva students: 'As long as my soul remains within me, I will not part with it.'[1] He could not be broken, and the unshakeable presence of his faith provided an example of endurance until his death in November 1943. Even though the Jewish community faced annihilation, these resistors survived the attack on spirit and never relinquished an emotional and psychological attachment to an identity sustained by ancient texts and nurtured by cultural practices. That alone, the survival of faith embedded in words and practice, and the sustaining of a tradition in time, demonstrated the power of violent *and* spiritual resistance, and the courage of the men and women who fought with body, spirit and will.

Yet the stories also bring with them uncomfortable moral positions and ethical interpretations. As one survivor put it to me: 'We made up our own ethics in the forests, since the old ethics only meant death.' Much in partisan action involved revenge and retaliation against the murder of defenseless Jewish men, women and children. Partisans had to kill; make decisions over whether to let infants live or die; whether to admit unarmed escapees from local ghettos into their partisan units. Much of the compassion exhibited by the writing of Jewish law during that period concerned impossible moral dilemmas in a community facing annihilation. One rabbi, for example, was asked by a young Jewish woman whether Jewish law would allow her to become the mistress of a Polish man because, by doing so, she could save her life. And there were the women who inadvertently smothered their infants while hiding from roundups who asked rabbis whether God would forgive them for such a tragedy.

Rabbinical preaching develops against a backdrop of increasing psychological and physical breakdown in the ghettos. How are we to understand Rabbi Shapira's insistence on the absolute character of faith in an environment where children die on the streets; where starvation, illness and madness claim hundreds of lives each day; where entire families are transported to Treblinka and Sobibor?[2] What is the ethical and practical meaning of 'faith' in a world where corpses seem to be more numerous than the living?[3] It is questions like these and their understanding in the context of spiritual and violent resistance that constitute the focus of this book. I want to look at the partisan point of view; but to interpret it through choices they found difficult to make; and the kinds of demands which a murderous environment placed on the partisans' evolving, new moral positions.[4] Similarly with the spiritual resistor: what psychological and moral space was created by spiritual resistance and how might we see that space, in historical retrospect, as an act of negation, in effect saying to the enemy, 'You may massacre my body, but I give you nothing of my soul'? Or as it was put by a Hasidic Jew in a small Polish ghetto: 'They can take my body – but not my soul! Over my soul they have no dominion!'[5] Was spiritual resistance to be admired or was it, as many partisan survivors described it, 'futile gestures, with not even the hope of saving lives'? Lastly, and very simply, what I wish to convey in this book is that during the Holocaust, the Jews, in a number of different venues, mounted significant resistance; and that knowledge of such resistance should put an end to the all too common belief that the Jews did nothing to resist their own fate.

1
The Moral Justification for Killing

He sits across from me, in a nondescript Manhattan hotel coffee shop, but his enthusiasm fills the room. What a likeable guy is Zvi Bielski, the son of a leader of one of the most successful Jewish resistance groups. The Bielski Unit, operating in the Byelorussian forests, composed of 1,200 Jews, 300 of them fighters and the rest a support community, survived virtually intact at the end of the war. The unit suffered just five percent casualties over a three-year period, an extraordinary achievement considering that most resistance units suffered over fifty percent casualties. Listening to Zvi describe his father, I find myself fascinated by Zush [Zeisal] Bielski, his courage and – from his son's perspective – his ferocity. It is almost as if Zush is sitting there. His son's admiration for his father is infectious; I am drawn to this courageous fighter, even though Zvi's narrative depicts him as a fearsome and often implacable man. Yet, as Zvi reminds me, 'Survival was at stake; you had to be merciless.' Zush and his fighters were merciless; they killed without guilt, without hesitation. They killed Germans, collaborators and, in a few instances, families of collaborators, to drive home the point that Jews were not to be handed over or betrayed.

Zush died in 1995; his brother, Tuvia, passed away in 1988. Tuvia, in Zvi's account, comes across as a stirring, determined leader, a political and organizational genius, able to hold in check his impetuous brother and to negotiate with Soviet generals the status and supply of his unit. Yet, Zvi insisted in pointing out to me that Tuvia's political success, in the complex world of Soviet partisan politics and Russian anti-Semitism, depended on awareness on the part of the Russians that the Bielski Unit would kill to protect itself and its integrity. And if the Russians tried to disband the exclusively Jewish Bielski group, that would be met with violent resistance. Generally, the Soviets discouraged and did actively dismantle Jewish partisan units. In part this had to do with the Soviets' belief that no unit should have a 'religious' or 'nationalistic' identity; but much stemmed from the Soviet partisans' anti-Semitism and an unsympathetic and often hostile reaction to Jewish units.

Zvi Bielski minces no words about his father's position, and while he consistently pointed out to me that there are many interpretations of how the Bielskis operated, there is no doubt about the reason for their success: their willingness to use violence to defend the group, and to make the point that collaboration and betrayal would not be tolerated. 'My Dad often said that if he had fired more bullets, he might have saved more people.' Zush's first wife and baby had been killed by the Germans; they never made it to the forests. His bitterness transformed into a mission with a dual purpose: to kill the enemy *and* to save as many Jews as possible.

> 'People were afraid of my father. The Bielskis, if they had to, would wipe out an entire village as an example not to kill Jews. Zush went into ghettos to recruit and told young people if they came with him they might have a chance at survival … make no mistake about it; these guys were vicious killers when they had to be … Bandits, but their violence was not indiscriminate; they killed only to protect themselves.'

Zush never operated in a gray area. No middle ground defined his behavior; it was either kill or be killed; the objective was to survive for the day, the week, the month. Zush and his fighters intimidated and threatened peasants for food, but never killed for food; their victims were those who gave away Jewish positions, spies sent by the Germans and any German unit that managed to penetrate the dense Byelorussian forests where the unit survived. Ferocity kept this band of Jews alive. 'My Dad said to me a million different ways, we needed to survive, but if they fucked with us, we would kill them… . I would kill at a moment's notice.'

When I asked Zvi about his father's attitude to spiritual resistance, which is also a subject of this book, he summed it up in a terse comment of his father's concerning religious Jews. '*Daven*ing [praying] for what? God is not going to show up here; what we need is guns; if there is a God, what the hell is he doing to us?' Zush's reaction expresses little ambivalence; it was, however, not the case that spiritual resistance in partisan and underground units disappeared altogether. As I will argue, it was there in subtle forms, but spiritual resistance possessed little meaning and significance for most resistance fighters.

Simon Trakinski, active in the Vilna underground and later with partisans in the forest, tells an interesting story about spiritual resistance. He had been in a labor group periodically taken outside the ghetto to work. A young Talmudic scholar worked beside him.

> 'Someone in the group asked this devout Jew, "Whom would you prefer, Stalin or Hitler?" He thought for a while and then said Stalin would be worse than Hitler. We were of course quite surprised by this; Hitler and the Germans murdered Jews every minute and every day, and Stalin and the Soviets fought. "How can you say that; look at what's happening," we said. But his response surprised me. "Look at what's really going on. Hitler will not succeed; those Jews who survive will come to life again and restore the Jewish community and be more Jewish, more devout than before. But under Stalin Jews will disappear; they will be assimilated; their religious artifacts will disappear; and if the Jews don't end up in Siberia, they will disappear into Russian culture and lose their sense of Jewish identity. Stalin wants to do away with Jewish cultural institutions and practices; he is therefore more dangerous. With Stalin's aim of breaking down Jewish identity, before long there will be no Jews at all; no culture, memory, remembrance and no prayer. *Shul*s and artifacts will be lost; bodies might remain, but their historical sense of Jewishness and their religious theology will be lost to time. Stalin therefore is more of a threat than Hitler, because with Stalin there will be no Jews left at all; the Jewish future is dead." For this guy, only identity mattered; Jewish identity had to be maintained at all costs. He even refused to shave off his beard and hid it in his jacket whenever Germans or locals were around. What a fanatic; but I admired him.'

Unlike Zush, Simon strongly believed in preserving Jewish customs and rituals, no matter how dangerous. Observance in the forests not only focused his Jewish identity, but emotionally kept him connected to his past.

> 'We tried to keep up practices; my mother baked matzoth from dark flour; in 1943 when we were fighting in the woods, we fasted on Yom Kippur. We were already very hungry, but we refused to eat and fasted the entire day. You see, we tried to live a normal life; and being an observant Jew meant being a normal Jew. We had to keep up "normality" even in the woods. Religious practice and theological study for most of us had been our normal daily life, particularly during the holidays. We weren't theologically correct Jews, but the culture of observance had been very much part of our normal life.'

Zvi, however, forcefully reiterated his father's position regarding prayer and theology: what saved lives was violent, armed action. It is true, faith could not save Jews from death. But the significance of spiritual resistance should not be ignored. Like the role of prayer and faith in the ghettos and for Jews heading for the gas chambers or the forest, killing needs to be understood as a form of psychological defense, a last-ditch effort to preserve the dignity of the spirit and the integrity of the self. It is a position, I will argue, to be admired and placed in the context of the brutality of German oppression and the genocidal policies of German institutions, including the SS, the army, industry and the professions, especially medicine and science.

But Zush Bielski, a genuinely brave and fierce man, would have none of the 'spiritual': to save lives meant engaging in acts that had nothing to do with spirit or observance. Killing required arms and a willingness to resort to violence that showed no mercy to the enemy.

> 'Both my Dad and Tuvia knew they were in the toilet bowl of the earth; and they had to get out of it. They were willing to kill, whoever got in their way... . it was both Tuvia's politics in holding the unit together and in dealing with the Soviets, and my Dad's hatred and toughness that allowed the unit to survive.

But the Bielskis were not just about killing; they were also about saving people.'

Listening to Zvi describing his father, it was hard not to be drawn to Zush Bielski and to think that if more Jews had been as fierce, maybe the slaughter would not have been as vicious or as genocidal. Yet, the story is much more complex; and it would be wrong to condemn Jews for not resisting more or to have expected more resistance than there was. I shall look at factors that blocked the possibilities for resistance. Much Jewish behavior had been conditioned by history and the culture of getting by and negotiating differences. But the suddenness of the German advance and the unremitting war the Germans waged against Jewish children and Jewish families through ghettoization and the policy of mass reprisals severely restricted the Jewish response. Yet, it is impossible, now, not to admire the courage of the fighters and the charisma of men like Zush Bielski.[1]

Zvi left me with a story of his father and Tuvia, both in their sixties, living in Brooklyn, and an image of determination that reflects where they came from:

'I had just bought a new Harley Davidson bike; I was a teenager really excited about riding this magnificent machine. I'm standing in front of my house about ready to start it up; my Dad and Tuvia are outside watching me. Suddenly my Dad says, "Get off that damn bike; I'm going to ride it." Tuvia is standing there laughing, saying to me, "That fucking guy, he can do anything." I don't know if either of them had ever ridden a bike before. So, Zush grabs the bike; Tuvia climbs on the back, and they both roar off down the street. Tuvia and Zush Bielski, heroes of the Holocaust, saviors of 1,200 Jews, careening down the street on the back of a Harley in Brooklyn. What a sight!'

Decisiveness, guts and courage appear in the narratives of resistance survivors. It is not that they are saying 'we were stronger than the others who died'; but through luck, and the bitter conviction they were doomed, they would rather die fighting. That desperation, what was required no matter the cost, the awareness that the Germans were imposing a universe full of death with no way out, forged an urgency that emerges in the power of their words.

After the initial German occupation of his village, not far from Lublin, Frank Blaichman wanders in the forest, lost, confused, as are all those he meets who managed to escape the German roundup. He is afraid: 'My insides are crying because I knew my parents, my family had been put on transports.' He speaks of Poles raping Jewish women, random killing of children and old people by both Poles and Germans; he organizes a unit who arm themselves initially with pitchforks and sticks. He extorts his first weapon by disguising his pitchfork as a rifle and intimidating a local peasant into turning over his guns.

Frank tells me he became a new man when he overcame his fear of being killed, when death no longer mattered, when the bitterness of his family's murder turned into hate. At that moment, he said he could act; his 'destiny was in our [fellow resistance fighters'] hands.' If refugee campsites had not been protected by resistance fighters, the Germans and local collaborators would have murdered everyone in them. 'If you had fear, you couldn't survive; so we did all we could to instill fear of us in the villagers; they knew we would execute collaborators.' When Frank's unit captured collaborators, they interrogated them, found out who had sent them to spy or had paid them for betraying positions and then killed them. 'Some of these guys begged for mercy; but we said to them, "What mercy did you give to our people?" After executing collaborators, the unit felt great ... they could no longer hurt us; it gave us a feeling of pride and satisfaction at knowing we could protect our people.' His unit even managed to hide a group of children in a bunker near the house of a farmer whom they threatened to kill if he gave away the children's position. 'We never sought big battles with the Germans; we did what we had to do, to protect our people. The villagers were afraid of both us and the Germans.'

Participation in these units transcended class and ideological barriers: 'We become one people, one class.' But there was also the conviction that what they did required a personal commitment to revenge. 'What I'm doing, I'm doing for myself, and that meant retaliation for the death of family and loved ones.' Again, like Zush Bielski, Frank set out a clearly defined moral universe for himself, without any gray areas: 'We were good to good people and bad to bad people; we created the respect necessary to make people into good people. So Polish and Russian partisans did not mess with our unit; they knew we would cut off their heads if they did.'

With Frank, as with the Bielskis, spiritual resistance played little or no role. 'With religious Jews it was "with God's help we will overcome this; and if God wants it another way, who are we to argue with him?" Religious Jews were locked into their beliefs; they had nowhere to go except faith.' Yet, like all accounts of the Holocaust, the practices of faith present complex moral issues bearing on survival; the going inwards, the power of belief, made a difference, not in the way that violent resistance saved lives, but as a psychological fortress that maintained identity for Jews not as abject slaves but as human beings who faced death with a part of the self untainted by German brutality. Yet, with few exceptions, Frank's attitude, unlike Simon's, mirrors resistance fighters' view of faith. 'Time did not permit us to be religious ... many lost faith; holy Jews got murdered. Our Jewish identity during that period was to try to save as many Jews as possible and to kill as many Germans and collaborators as possible. That did not permit us to be religious.' But for the Jews who never made it to the forests, who were murdered or starved to death in the ghettos, what did spiritual resistance mean? It is that question that needs to be examined with considerable sensitivity and awareness that for the vast majority of Jews during the Holocaust, the possibility of exit was foreclosed almost immediately after the German occupation.

The difficulty of exit for young Jews who wanted to fight cannot be overemphasized. As Simon describes it: 'There was no escape on the outside; on the outside we faced the Lithuanians and the Germans; all wanted to kill us.' The *Judenrat* (Jewish Council set up by the German occupiers) in most ghettos gave little assistance to those wanting to leave. The Germans sealed off the ghettos and made it clear that anyone trying to leave would be killed and the ghetto would be subject to mass reprisals and the murder of family and friends. The *Judenrat* were rightly terrified of mass reprisals and with few exceptions would have little to do with the underground units or partisans. Simon: 'We were fenced in like cattle in the Chicago stockyards. The *Judenrat* of course wanted us to stay because of the policy of mass reprisals. Germans if they caught partisans would kill their relatives, anyone they could get their hands on. But I believed that every honest man's place is to go into the woods.' But, he cautioned, that was 'easier said than done.' Simon was fortunate; he made it out with his brother: 'I was lucky.'

Contingency played an enormously important role in surviving. Although it would be wrong to attribute survival to chance alone, resistance fighters needed plenty of luck to survive the unexpected. Simon tells a story about an incident that occurred a few nights after he left the Vilna ghetto. He and a friend went to sleep; but Simon kept his boots on because they were serving as a pillow for his comrade and he didn't want to wake him. Suddenly, he awoke to gunshots. Immediately, he leapt to his feet and ran as fast as he could to cover. His friend, who slept with his boots off, never made cover because those who took the time to pull on their boots were killed. Simon escaped to the swamps with the help of Markov Brigade partisans.

Frank's and Simon's recollections convey extraordinary accounts of endurance and strength. But it was the look in their eyes that seemed to convey a memory that was absent from their words. Perhaps it was their sadness in knowing how important this history was to them, what a critical part of their lives it had been, and wanting it to be heard. Perhaps it was their unstated but very real sentiment that soon they would not be able to recount these experiences. Or perhaps they felt no one would understand fully what they had been through, what they had created and what they had endured to survive. Yet these same men seemed at home in their Manhattan world, content and satisfied with lives well lived. But maybe that contentment concealed a terrible pain, because always in their eyes lay that other home, the one that had vanished, the one that spoke of death and tragedy.

Simon remembers his mother, her bravery in trying to protect her family.

> 'We knew as early as the winter of 1941 what the Germans were up to; I was only seventeen years old when we found ourselves relocated to the Vilna ghetto. My mother at the time was only 39. When she could, she left the ghetto to find food; she wrapped a shawl around her shoulders to hide her Jewish star. She witnessed beating, executions by the roadside; everywhere she went she saw Jews being killed. No one was in the dark regarding German intentions.'

I listened to stories of victory, transcendence, violence and loss; but another story lay underneath these narratives, the truncated feelings of children losing parents, sisters, brothers, grandparents, and the security of a childhood thrown into chaos. It is that story that never leaves the eyes, even as Frank looks at me, silently for a minute or two, while I wait for the elevator. It is a silent communication – of fury and of immense sadness at a part of his life he knows was lost forever in those forests. One imagines that by the time we reach old age, certain compromises have been made with life. But with survivors like Frank and Simon, firm in their support of Israel and George W. Bush's position on Iraq, whatever peace or compromise lies in this life offers utterly no compensation for the peace lost then; for the criminal ripping of children from their parents and the unbearable grief of knowing your parents and those you love are dead but not knowing where, of not being able to find a grave on which to place a stone. Perhaps home, in the unvoiced message of their eyes, had to do with the past and present absorbing each other, emotions as alive now as they were then, the past never leaving, being here and there, an immediacy conveyed through silences and staring through the window of an elegant living room to spaces far beyond the winter cold of a Fifth Avenue park.

In their recounting, memories possessed immanence, stories full of dilemma, guilt and regret. It was most strongly put to me by Simon at the very beginning of our interview: 'Whenever I go to a funeral now I feel envious; I never had the privilege of burying my own parents, of saying *Kaddish* at their graves.' Being there, with their memories, wanting to tell their stories, remembering the pain and isolation, moving into it without embarrassment, and the sadness in not being able to bring back the ones they loved – this is the irretrievable part of their histories. Perhaps it is this remembrance that lay deep in their eyes, knowing that their parents had not been buried but had been murdered in a dreary village, forest or camp, with no opportunity for these men to say *Kaddish*. Or perhaps a nagging doubt or a moral lapse lies close to the surface. Simon:

> 'It's hard for me to tell you this story. It's as real now as it was then. We had been chased by the Germans and their collaborators into the swamps. We found a small island with a few huts

on it; we hid there for several days. In the hut next to us, we heard the cries of a baby, really a kind of whimpering because it was so weak. The Jewish "wife" of a Soviet commander had given birth to their kid in that hut but her husband insisted she leave the baby. A baby's crying couldn't be controlled, so infants presented real dangers to all of us. If the enemy heard the baby, he could find us. We listened to the whimpering for eight days; the baby died on the ninth day. His mother visited him once. Life was not only cheap, it was incidental; the mother moved only a kilometer or two from her child. But she refused to take the baby with her. I imagine she couldn't bring herself to kill it; so she just let it wither away ... and we did the same thing: we refused to save the baby, because we knew the danger. But, we suffered that baby's death: it was not easy listening all those days to the cries of a dying infant. You see, it was the spirit of the times; life had become so cheap; you could lose it anytime ... the priority was survival.'

Simon seems to be looking far beyond the walls of his apartment as he tells me this: 'We had no place to go, what were we to do? Take the baby with us? You couldn't do anything: neighbors who lived next to you for years would hand you over to the Gestapo for a loaf of bread; human sensitivity disappeared. I couldn't afford to feel anything for that baby; the baby threatened me; in the end I had to choose my own life and those of my comrades.'

Yet, these are not unhappy men. Frank and Simon seemed to me to be full of life and enthusiasm; sharp and incisive in their observations, engaged with the contemporary world, healthy, and with humor and endless tolerance for my questions. They were proud of their successes in the United States, their children's accomplishments; and they brought out snapshots of their families back then and now, of groups of fighters in the forests. In their survival they seemed to say 'we were lucky,' because they all spoke of luck. But I also sensed another message; 'maybe we were not so lucky – we lost our entire families and we can never bring them back. We had to make terrible choices. And when we speak to you, we remember all these choices, and the tears are as fresh now as they were then.' Maybe that's the sadness I felt leaving these interviews, the sense that the past is not over; that these political resistors carry inside them a set of moral perspectives for the present, for us who were not there. And in recounting these numerous acts of courage and tough choices, they ask the audience to listen and not to judge. They want us to know something of what it meant to be there in those barren fields, prison camps and dense forests. What these men and women accomplished is political in the most profound sense of the word: the undergrounds and partisans preserved the political space of identity and freedom, creating communities of friendship in primitive forests and enduring the most unimaginable hardships. The public life of these fighters and resistance groups and underground organizations is not the public space of institutions, but these fighters and survivors created public spaces carved out of desperate times, and whose very existence contributed to the survival of thousands who would otherwise have died.

No moral ambivalence framed the narrative of Miles Lerman; the German assault rendered traditional moralities obsolete and dangerous to survival.

> 'The peasants eventually took us seriously; we had no hesitation – we would kill whom we had to. If we had to burn a peasant village to protect ourselves or punish an informer, we would do it. The Germans looked at us like we were mice or rats; they would trade sugar or vodka for Jews. Some peasants gave up Jews for a bottle of vodka. This happened in our area; a peasant had trapped a couple of Jews by offering them some food, and then turned them over to the Germans. One night we showed up at the house of the peasant and hanged him and put a sign on him that said, "this will happen to peasants who betray Jews". If we had to, we would kill an entire family; there was no other way to protect ourselves. If the peasants would hand in Jews for a glass of vodka, now, how do you handle that?'

The ghetto: demoralization and breakdown

Critics of Jewish inaction, like Hannah Arendt and Raul Hilberg, capture an important reason for the absence of a more violent mass resistance:[2] the role of the *Judenrate* in collaborating with the Germans. Yet, even that story should be treated very carefully, since members of the many *Judenrate* in both the large and small ghettos believed that cooperation would assure the survival at least of a remnant of the Jewish community. A physician whom I interviewed in Warsaw, who had been in the Lodz ghetto, told me that after the war he would have been first in line to kill Chaim Rumkowski, the notorious head of the Lodz *Judenrat* who continually bartered away Jews for selection. 'But now I regard him as a great man.' I was surprised by this since Rumkowski facilitated the infamous exchange of children under the age of ten and the elderly over 65 for several thousand Jews who were capable of work. 'You ask me why I think he is now a great man? Because he kept Jews alive in the ghetto longer than any other ghetto leader.' In the spring of 1944 some 60,000 Jews remained in Lodz, until they were all transported and murdered in Auschwitz later that summer and fall.

While in retrospect that strategy was fatal, Rumkowski appears at least to this survivor to have engineered strategies that prolonged survival – of at least a remnant. The point Dr. M. was making was that moral culpability is difficult to assign, and we should be very careful how we evaluate terrible decisions imposed on the Jewish community, although in *Eichmann in Jerusalem,* Arendt expressed moral outrage at *Judenrat* collaborators.

At what point should the *Judenrate* have realized the enormity of German intent? The ordinary, compliant men chosen by the Germans for these leadership positions could not be expected to transcend their view of survival and the lies continually fed them by the Germans. Spiritual and political leaders that might have generated resistance had been executed soon after the Germans occupied Jewish villages and towns. Many radical political leaders escaped to the Soviet Union after the German invasion of Poland; some returned to participate in and organize armed resistance. Leaders of stature in the traditional communities were quickly murdered, including many rabbis, long before the German authority appointed the *Judenrate*. It is also the case that the instinct for survival, although understood as collusion, molded *Judenrat* policy.

Many joined *Judenrat* administration and the Jewish police to assure their own survival and that of their families. Many believed that to be a member of a Jewish police unit would be a shield against German roundups, or that work in a *Judenrat* office would make it less likely that they would be placed on a selection list. The human self when faced with terror reacts with terror. The *Judenrate* operated in an environment of terror, and those that worked for the *Judenrate* acted in ways they believed would save them and the remnant. One can fault their tactics, but given the realities of human nature, could they have been expected to act in any other way? While their policy of collaboration failed miserably, it possessed both a strategic and a moral logic.

To violently resist the Germans, to take an active stance against *Judenrat* policy, meant the resistor self had to transcend its own terror, fear and uncertainty, to see the possibility in alternative forms of political and social organization. To their credit, the *Judenrate* in most ghettos, with the cooperation of social service groups, sponsored and supported hundreds of soup kitchens and fed thousands of homeless children and refugees, the sick and elderly.[3] In an environment where hopelessness defined everyday life for hundreds of thousands, the establishment of the soup kitchens, at least in the short term, inhibited the German policy of mass starvation. By the time, however, that individual *Judenrat* members realized the full extent of German policy, it was too late; in Warsaw in the summer of 1942, hundreds of thousands of Jews were murdered in the death camps of Treblinka and Sobibor, in addition to the tens of thousands dying between 1939 and 1942 from disease and starvation.[4]

But let us assume, for the sake of argument, that the *Judenrate* were headed by resistance leaders; that each ghetto *Judenrat* had been the center of underground activity. Would the outcome have been any different? Probably not, and for one very good reason. Family, in addition to religious practice and its organization, had been central to the cultural and public life of the Jewish communities in Europe. The Germans understood that. A centerpiece of their policy lay in an assault on the family – in particular through the killing of children – and on rabbis and Jewish sacred objects. Demoralization of family life, the desecration of religious artifacts and the despair provoked by the selection and killing of children are critical to understanding why so many Jews were murdered. With some exceptions – for example, the Jews of Budapest, who were transported directly from Budapest to Auschwitz in 1944 – most East European Jews had been driven from their villages and home towns into ghettos, if they had not already been slaughtered in the process.

The route to the death camps for Jews from France, Belgium and the Netherlands was equally defined by rapid confusion. Jews in the Netherlands, for example, had to endure Westerbork, a massive holding camp, before being transported to Auschwitz.[5] The suddenness of the German assault on the primarily middle-class West European Jewish community meant that families had to face the Auschwitz journey alone, confused, hungry, sick, without any chance for political organization, wondering where they were going. Some Jews from the West were transported to Auschwitz in comfortable railway carriages, having been told by the Germans that at their destination they would be housed in hotels and the men assigned jobs. Some ended up in the ghettos of the East. Children often were separated from their parents and taken away to special blocks and demarcation points. Some families, like the Franks, hid in cellars, attics or specially constructed rooms. Families fought with each other; strangers often were forced to live in the same room. The inability to protect or rescue children and infants had a devastating impact on consciousness and will, as it would on anyone unable to protect their family from violence.

Demoralization produced by ghettoization is recorded in countless diaries describing life in Vilna, Lvov, Bialystok, Theresienstadt, Kovno, Lodz and Warsaw.[6] What emerges from these diaries – many written by teenagers with a painfully clear grasp of family life – is a picture of a community devastated by barbarism, never embracing underground movements, partisan fighters or strategies based on violence that might bring mass reprisal. Diary entries critical to an understanding of the Jewish community's despair describe a population increasingly suffering physical weakness and debilitation, and forced to confront the terror of not being able to protect children. By the time these victims reached the gas chambers – if they had not been killed by disease, brutality and mass starvation – they had been depleted by hunger and the war waged by the Germans on physical survival.

It is that story – the progressive and unrelenting German assault in the ghettos – that needs to be examined *side-by-side* with Jewish violent resistance and partisan action. The one cannot be understood without the other, since out of the psychological environment of despair came the ghetto undergrounds and partisans, the fighting units. The resistance faced tremendous odds from the

Germans and the despair in the ghettos, although many fighters found moral comfort in the violence of resistance. Lerman:

> 'We weren't heroes; I'll tell you about a hero, a little girl with us, about twelve years old. She had blond hair and blue eyes and looked very Polish. So we would send her on courier missions, to get supplies, send messages, things like that. For medicine we were always in need of iodine; this was before penicillin and iodine was effective in treating a number of different infections. We sent her into a village for a couple of liters of iodine, but she was caught. Someone told us the Germans offered to send her back to her mother if she would talk; if she would tell where our camp was. But she never talked. They tortured her horribly; but she died never having revealed our position. That child, she is the hero. They never intended to return her to her mother; by then her mother probably was dead.'

The most dramatic underground action – the Warsaw Uprising of spring 1943, when only 50,000 people remained in the ghetto – gathered support from at the most a few hundred fighters.[7]

The devastation produced by the massive transports of the previous spring and summer had depleted the community of the sense of itself as a world with a future; life in the ghetto had been reduced to a monumental effort simply to survive physically, with many pacing the empty streets wondering when their turn would come.[8] The survivors of previous roundups watched as family members disappeared and died. Diaries of the mass roundups in the summer of 1942 describe children and infants killed or taken to the central train station, forced to wait for days without food or water; parents returning home to find their children missing; children coming home to find their parents victims of roundups. The descriptions of death and dying on the streets, so graphically represented by Roman Polanski's film, *The Pianist*, the epidemics that ravaged the ghettos claiming thousands of lives; the absence of food and sanitation, and the processions of death wagons daily, hauled often by children, through the ghetto streets: to read these diary accounts is to witness a *physical* universe in the process of disintegration.

The 50,000 left in Warsaw each suffered personally, if not the loss of a child, then the murder of a father, mother or grandparent. No one remained who had not been emotionally devastated by the effects of the transports. When the Germans attacked the ghetto in March 1943, in the final phase of their plan to kill every Jew remaining in Warsaw, the community had little or no resources to support the few hundred fighters. These people could barely survive themselves, and the killing by the Germans that accompanied the uprising decimated the remaining 50,000.

But underground resistance occurred in a physical and psychological universe where Jews fought back. Lerman: 'Many understood the power of starvation and many who starved fought back.' Even in Vilna, Warsaw, Kovno and Bialystok,[9] where the underground distributed thousands of leaflets and posters describing the Germans' intent, the recruitment of fighters faced tremendous obstacles, not the least of which was the fear of the breakup of the family. The attack on family bonds, the war on children, inequality within the ghettos, hierarchies based on proximity to the German command, the possession of labor cards as opposed to their absence, access to food coupons as opposed to little or no access, all were significant factors contributing to German efforts to immobilize the possibility of resistance. Yet, even with the terrible moral choices the resistance had to make, survival hinged on consciousness liberating itself from ghetto mentality and from traditional moral understandings. Lerman again:

> 'When any of our women had babies, it was an unspoken law the baby would be strangled and killed at the moment of birth. It would have been impossible to have had a baby in those circumstances; they could not be allowed to live. No one talked about it, but it was painful. Perhaps we shouldn't have killed them; but who knows: even one cry could have given away our positions.
>
> We were in the forests from 1942 until June 1944 when we were liberated by the Russian army. Perhaps 70–80 percent of our unit survived; if our group hadn't been in the forest, 90 percent would have perished. Were we lucky? Of course; there was good luck and bad luck; we had good luck. But being partisans we were more likely to have good luck; we could fight, unlike those in the ghetto who all perished. But I could have been one who died; the Germans would burn buildings, round up people randomly and

kill them. We all shared the same bitterness, and I have no apologies for my hatred.'

Morality in the forests had to be drastically revised; survival depended on discarding old moral beliefs. Liberation of self meant, as well, creating military vengeance and rewriting the laws of community.

What, then, I want to do in the following chapters is to describe and interpret significant moral and ethical positions guiding this effort at resistance. But I will look at the dilemmas and conflicts in resistance as they affected both violent and spiritual resistance. Neither path was an easy one to take; each involved significant moral demands and faith in choices which ultimately meant the difference between life and death and the choice about how one died. The violent resistor had to literally relearn moral positions, creating an ethics adaptive to the demands of survival; the spiritual resistor fought against the ever-present reality of madness and the sinking into apathy. For both, resistance preserved sanity and protected the self's integrity from the implosive power of genocidal action.

2
Collective Trauma: The Disintegration of Ethics

In Budapest 1944, during the massive German project spearheaded by Adolf Eichmann to exterminate all Hungarian Jews, a group of young Zionists rescued thousands of Jews from the prospect of transport to Auschwitz. Issuing fake identity papers, impersonating German officers and Hungarian Iron Cross troops, finding and establishing safe houses, these young people never relented in their efforts to thwart German plans.[1]

In Nesvizh, Poland, a group of Zionist youths formed an underground and began acquiring arms. What stood between this group and their objectives was the *Judenrat*. One of the underground leaders writes:

> 'In addition to reorganizing the resistance to strengthen the ghetto community for revolution, it was vital to undermine Maghalief's position as chairman of the *Judenrat*. His methods were unscrupulous. He believed that bribery alone would ward off calamity, and since he was our only representative to the German command, he had to be stripped of his power.'[2]

In Nesvizh, the sentiment of resistance countered the *Judenrat*'s approach to appeasing the Germans and selecting Jews for transport. Several ghetto inhabitants inspired by the resistance armed themselves and launched a putative attack against German troops. The German army, with the help of local police, quickly subdued the insurrection. With the ghetto in chaos, many escaped into the surrounding forests. The isolated, but free ghetto inhabitants sought out partisan units.

Resistors in the ghetto faced enormous odds.

'So, during the winter of 1942–43, we basically lived like squirrels, hiding in a hole. As you can imagine, the air in the bunker stank like hell. On nights when it was snowing or otherwise very dark, we would lift open the cover of the bunker a bit. Otherwise, we would sit in our hole ... no one in our group expected to come out alive from that hell. The main thing was not to be taken alive by the Germans, not to submit to their questions, their torture, and a passive death at their hands. We were always armed and had an understanding that if we were ambushed, we would fight until we were killed. If need be, we would shoot one another rather than be captured. It was inevitable that we would die – but death would come on our terms.

Once I became used to that idea, I became extremely brave.'[3]

Nowhere could the desperate condition of the ghetto be better witnessed than in the ghetto hospitals. Adina Blady Szwajger, a survivor and a doctor who worked in ghetto children's hospitals in Warsaw, describes conditions in what can only loosely be described as 'hospitals': 'A not uncommon sight: children appearing at the hospital with their heads and bodies covered with lice'. One child's head '"looked grey". It was only when you came up close that you could see that the grey mop of hair was moving.' Ghetto desolation and a loss of will permeate another child's description of his family's fate: 'When my sister died, Papa said it didn't matter where they buried her because we wouldn't live to visit her grave anyway. And Papa wrapped her up in paper and took her out into the street.'[4] What it feels like to wrap the corpse of your child in paper and deposit her on the street is unimaginable, an act existing in a universe far away from the present. Yet, no huge imaginative leap is required to surmise that such daily occurrences had an enormously depressing impact on the ghetto populations. Homeless, lice-ridden children running through the streets snatching pieces of bread, begging, knocking at doors and dying in the gutter, brought to the ghetto a paralyzing hopelessness. During one German action, Dr. Szwajger administered lethal doses of morphine to infants to keep them out of German hands. Never enough food to satisfy the hunger of sick and dying children, any scraps that became available were viciously fought over.

> 'One day, on the older children's ward, the famished skeletons threw themselves at the soup pot, overturned it as they pushed the nurse away, then lapped up the spilt slops from the floor, tearing bits of rotten swede [from the floor] away from each other.'[5]

Vomit and human waste fouled the hospital wards. The children barely resembled human beings.

> 'On the bunks lay skeletons of children or swollen lumps. Only their eyes were alive. Until you've seen such eyes, the face of a starving child with its gaping black hole for a mouth and its wrinkled, parchment-like skin, you don't know what life can be like... . There weren't enough mattresses for the bunks and their number was diminishing because the bloody diarrhea reduced them to pulp.'[6]

Dr. Szwajger describes a child 'who, in the frost, had stripped naked on the street in order to be taken in by the hospital because children came to us for that one last thing we could offer – the mercy of a quiet death.'[7] Scenes like the following were commonplace:

> 'When my sister died and Mamma carried her out, she didn't have any strength left to go and beg, so she just lay there and cried a bit. But I didn't have any strength to go out either, so Mamma died too, and I wanted to live so terribly much and I prayed like Papa did before, before they killed him.'[8]

A little boy, shot in the liver while running from a German policeman, hid a small coin in his hand (the equivalent of a nickel or a few pence) and said to the doctor before he died that she should give it to his mother. A six-year-old girl, who had witnessed the death of her brother, sister and mother, had the shuffle of an 'old woman.' She pleaded with Dr. Szwajger to let her remain in the hospital; with her parents dead, she had nowhere to go. The doctor speaks of one little girl who kept apologizing for the smell from her gangrenous legs. Children who were forced to hide day and night in small ramped holes or bunkers, or behind walls and in attics and cellars, developed rickets; they lost the ability to walk and, in some instances, how to talk because silence was essential to avoid giving themselves away to Germans, local police or informants. Births brought not joy but despair and fear.[9]

To escape from the ghetto meant hiding or trying to find sympathizers in the local population; but the Germans made it clear that the punishment for concealing Jews was death for anyone caught in such 'subversive' actions. Women who escaped to the forests faced the prospect of being raped by non-Jewish partisans or local peasants. Children had little chance of surviving the harsh conditions of life outside the ghetto; and while many in the ghetto knew of partisan bands and units where they might receive protection, most ghetto inhabitants chose to stay in the ghetto – not out of a death wish or apathy, but because they knew what it was like 'out there'; to survive in the forests required skills, endurance and logistics unavailable to the mass of the Jewish population.

Hiding outside the ghetto brought enormous risks. One survivor recalls: 'We couldn't relieve ourselves outside, because any farmers passing through the woods would have noticed the human waste.'[10] At any moment, those fortunate enough to build a bunker outside the ghetto could be discovered by German units, hostile locals or terrified farmers in fear that if Germans found Jews on their property, they too would be killed. Those escaping the ghetto faced the problem of food and supplies; of protecting against the cold, disease and injury. It was therefore terrifying to leave the ghetto; at least in the ghetto one had one's family – or what remained of the family – and friends after periodic selections. It took enormous courage to leave.

Despair inevitably preceded rage. 'Nothing mattered to me. I felt depressed and full of grief ...'[11] – this from a partisan who learned that her entire family had been executed.

Youth with a strong political belief, whether it was communism, Zionism or revisionist-Zionism, initially joined the undergrounds and promoted the complex and difficult process of trying to enlist ghetto support for underground actions. While there were notable exceptions (for example, Kovno), it was generally the case that the *Judenrate* feared the possibility of armed action within the ghetto, and this for a very good reason. The Germans made clear that reprisals would result in the execution not only of the underground fighter, but of the family of the underground fighter and whomever else the Germans chose to kill – friends, relatives, people at random. Mass reprisal for underground and partisan activity was an integral part of German policy, and it worked. The Germans had no compunction about killing scores of individuals to avenge the death of one German soldier. And even a show of force by the underground provoked mass reprisals. Yet the *Judenrate* on occasion would split about whether or not to aid resistance fighters; some members might smuggle supplies, money or weapons to the underground, while other *Judenrate* and administration members violently argued against offering support.

Death had as much presence in the ghetto as life. One Polish observer writes: 'People walking on the street are so used to seeing corpses on the sidewalks that they pass by without any emotion' (October 27, 1942). Another entry, of October 28, 1942: 'I saw an old Jewish woman unable to walk anymore. A Gestapo man shot her once, but she was still alive; so he shot her again, then left. People see this now as a daily event and rarely react.... It is a common occurrence that Jews come on their own to the gendarme post and ask to be shot.'[12] In dealing with efforts to mobilize the ghetto, the underground continually faced this ever-present desolation – the disintegration of spirit and will, particularly amongst refugees. Underground leaders like Mordechai Tannenbaum (Bialystok), Abba Kovner (Vilna), Chaim Yellin (Kovno), Mordechai Anielewicz and Yitzhak Zukerman (Warsaw) distributed leaflets and entreaties to a population shattered by starvation and the death of children. And in the days and weeks following large-scale *Aktionen* (roundups, deportations, mass executions), many believed that this disaster had to be the last; nothing more could touch the ghetto. Normality in the ghetto 'meant having just enough food to exist ... it meant the survival of the community while individuals were shot. It meant life behind barbed wires, like criminals, like slave laborers, without rest or relaxation.'[13]

Underground organizations generally were a loose association of various political groups; for example, in Warsaw, after the deportations of the summer of 1942, the Jewish Fighting Organization (JFO) was composed of Zionists, socialists, the Zionist youth, the Bund and the communists. Militant, anti-socialist, pro-violence Zionist revisionists, on their own, formed the Jewish Military Association. Both underground groups maintained contact with each other, and after the 1942 deportations, ideological and political distinctions made utterly no difference. Emmanuel Ringelblum, the Warsaw historian who organized extensive efforts at describing the social, cultural and political life in the ghetto, offers an explanation for Jewish inaction. It is worth quoting at length because his perception has received wide currency as describing the Jewish *group* state of mind:

> The Jews did not rise up against the slaughter anywhere; they went to their deaths without resisting. They did this in order that the others might live, for every Jew knew that lifting a hand against a German meant endangering Jews in another city or possibly another country ... to be passive, not to raise a hand against the Germans, was the quiet heroism for the plain, average Jew. It would seem that this was the silent instinct to survive of the masses, and it dictated to everyone, as though through a consensus, to behave in a certain way. And it appears to me that no explanation or exhortation would have helped – one cannot fight an instinct of the masses, one can only bow to it.[14]

As much as we can admire Ringelblum's work and the massive effort of the *Oneg Shabbath* in chronicling the decline of the Warsaw ghetto and Poland's Jews, these observations must be tempered by looking at some powerful psychological facts. Very little has been written on that amorphous Jewish 'mass' that Ringelblum and other diarists and chroniclers of life in the ghetto call 'the Jews.' By 'the Jews' is meant the vast majority of the Jewish population who never participated in organized individual political resistance, and for whom there is little record of 'spiritual resistance' in most of the six million who died, including the 1,200,000 children. The great majority of those murdered were not members of a political organization or Zionist youth movement like Hechalutz, Hashomer Hatzair, and D'ror; yet the psychological damage to the individual and the collective pushed the community into a numbed passivity. One survivor, Masha S., of the Kovno ghetto, remembers:

> Everyone seemed to be dying slowly; the resistance leaders would come into the ghetto and speak to us, but no one wanted

to do anything; just getting up in the morning took an enormous effort; no one had enough to eat; we didn't want to hear about resistance, we wanted to hear about getting more food; our rabbis had been tortured and killed; my mother and sister had been taken away [to be executed in mass graves outside the ghetto]. We had no will to do anything; I wanted to go with Yellin [the Kovno underground leader], but my father forbade it. It was my impression that Yellin had not much success with the Jews of Kovno. Don't get me wrong; it's not that we did not appreciate the resistance, although many feared it would bring reprisals. Very few wanted or even had the strength to get involved.[15]

Apart from the courage of the rabbis, faith and theology stood in the way of the underground's effort to recruit. The underground had little patience for prayer and spiritual sacrifice as substitutes for action. Rabbi Mendel of Pabianice:

'Hear, O brethren. Do you know where we are going? We are setting out in the holy road of *Kiddush a-Shem*. Our father Abraham led one of his seed to the sacrifice, and the two of them went joyously and enthusiastically. It is written: "And they went *yahat* [together]" – that is, they went with *hedva* [joy]. And as we all of us go together to the sacrifice – fathers with their sons, grandfathers with their grandsons, mothers and grandmothers with their grandchildren – What a great moment this is! Consider for a moment, dear Jews, holy and pure ones: Why are we being led to the sacrifice? Satan himself has openly and expressly said that he had found only one fault with us – the name Israel. Why, we are all going to sanctify the Name. But not too hastily, Jews. Rest a moment. These are our last moments.'[16]

While spiritual resistance was certainly heartbreaking in its dedication and faith, the majority of reports on the state of mind of Jews heading for the gas chambers or burial pits outside the ghettos describe human beings broken in spirit and will. Resistance fighters could not resurrect will where there was utter desolation. Resistance diaries, while critical of theological quiescence, demonstrate a strong Jewish identity; and while many political resistors practiced

Jewish rituals, most rejected religious sentiments that relied on God's will for salvation.

Chaika Grossman, a resistance survivor of Bialystok, writes: 'We did not develop any ideology of dying.'[17] Grossman describes the state of mind underground fighters faced in their effort to generate ghetto support. After a ghetto action in Bialystok:

> All the victims had not yet been found; they were still looking in the cellars and attics, still finding children who had been strangled when they cried in the hiding places, their mothers' own hands stopping their cries in their throats. People were running through the hiding places ... families had been butchered... . [Many] had gone out of their minds. [There were] cases of hysteria and madness, of shouting despair.[18]

For Grossman, antidotes against despair consisted of holding fast 'against depression and the feeling of impotence against a victorious [German] enemy.' Resist the Germans' attack on 'the limits of reality', and 'maintain action, discipline and the yoke of patience.'[19] Yet, the enemy had the power to alter reality, to destroy discipline and shatter patience. 'Perhaps [the members of the resistance] had not fully understood the agony of parents looking at their famished children. What use was there in living such a life?'[20] Ghetto conditions, the suddenness of the actions, the increasing, pervasive sense of desolation, made it impossible for underground fighters to convince the Jews that resistance indeed held out hope. 'We had no masses behind us.'[21]

Grossman blames the *Judenrat* in Bialystok. She argues that they knew the truth about the mass executions and that *Judenrat* members had heard reports about Ponary (the execution site outside Vilna) and the death camp at Treblinka as early as December 1941: 'Barash [the head of the *Judenrat*] heard the truth from me. He knew about Treblinka, and he himself testified in his speech of October 11, 1942... . everybody knows what happened in Warsaw, in Slonin ... [It is] better to go to Volkovysk [a nearby ghetto] than to Treblinka ... . It is not a lack of information we have here but a policy that was the direct opposite of that adopted by the underground.'[22]

The undergrounds continually fought to maintain violent resistance. But others in the ghetto fought to sustain spiritual resistance. For example, in the Kovno ghetto following the selection on October 28, 1941 when 10,000 of the Jewish population were taken to the execution site at the Ninth Fort just outside the city, a workman asks Rabbi Oshry, one of the few rabbis still alive in Kovno: 'Is there no obligation to study Torah before one dies? All along we have been studying, and now – no more?' Rabbi Oshry responds: 'I assured him that we would continue our studies.'[23] While this sentiment is admirable, it provided little comfort to underground fighters desperately trying to rally the ghetto into more forceful action and a sense of urgency about its fate. One wonders about the accuracy, much less the content, of Rabbi Oshry's observation that after the mass execution, 'Jews put even more effort into strengthening their trust in G–d Who would not forsake Jewry. Their prayers were recited with increased intensity and they studied Torah ever more seriously.'[24]

Increased study of Torah as a fitting response to mass murder had great currency in the Kovno ghetto. Rabbi Oshry writes: 'Extraordinary is the power of the holy Jewish Book. It has a soul of its own and encompasses the illumination of generations – a blinding light for those who would harm it.'[25] Children, he notes, hid their own *Chumash* [volume of Torah] or a *gemorah* [prayer or study book] and in doing so, took a great risk. 'I told them how dangerous it was to hide Jewish books. To this they responded, "Rebbi, if they shoot us together with our *gemoros*, at least we'll be sanctifying G–d."'[26]

Here is a clear instance of German ineffectiveness in breaking spirit, but even with these stories circulating in the ghetto, they did little to encourage violent resistance. A resistance fighter, who heard similar stories about children sustaining faith and continuing to pray, told me, 'I could only weep for those kids.'

Even such proscribed acts as the baking of *matzoh* for Passover did not constitute effective action in the underground's universe; while as Jews the underground respected such acts, what mattered more were weapons, people willing to shoot Germans and collaborators, and those willing to hide and support underground participants. From August 15, 1941 through the liberation of Kovno, Rabbi Oshry studied Torah with children. Masha S. again:

> In the ghetto we had to face rising prices, goods in short supply, work hours from dawn to dusk; no heat, or electricity; water that

made us sick; our limits were the limits of our bodies. We felt so terrible, physically and emotionally: and then a resistance fighter would hand us a paper; do you think we had time for that? It was all my parents could do to try to save the family; then my mother died, then my sister. All I had left was my father; we vowed never to leave each other.[27]

Nor could the undergrounds trust the local populations. While many Poles and Lithuanians expressed sympathy for the Jews' plight, when it came to concrete actions, the vast majority were intimidated by German orders or concurred with German anti-Semitic theories, or had been bred in local superstitions regarding Jewish evil. Here is just one example of this 'turning a blind eye'.

The German administration in a district outside Warsaw orders all hospitals not to give aid to Jews. Kowalski, who heads a local hospital, is concerned about this order; he calls the local police and is told to mind his own business. He posts guards at his hospital entrance and turns Jews away. 'There is no way I would have been able to save Jews, and certainly I would have been arrested and executed.'[28] In the next few paragraphs he describes how Jews are being randomly killed, how sorry he is, but how constrained his actions are by German orders. 'With my eyes I can still see the wagons filled with the dead, one Jewish woman walking along with her dead child in her arms, and many wounded lying on the sidewalks across from my hospital.' He describes Polish brutality against Jews and is particularly struck by the German order for the local *Judenrate* to pay '2,000 zloty and 3 pounds of coffee for the ammunition used to kill Jews.'[29]

Kowalski describes a situation where brutality had become so commonplace that bystanders were no longer affected by the pain the Jews suffered. 'Now when people meet on the streets the normal way of greeting is, "Who was arrested? How many Jews were killed last night? Who was robbed?" These events are so common that, really, no one seems to care. Slowly you become accustomed to everything.'[30] Death for the Jews evolved into the normal way of life; to see a dead Jew on the street was nothing out of the ordinary.

The underground had little success in mobilizing sympathizers on the outside. Underground fighters were constantly in danger of being turned in by locals who in return would receive a bounty from the Germans. A Polish academic I interviewed in Warsaw recalls the atmosphere of anti-Semitism: 'I was young, but I remember my parents and their friends saying things like "Well, at least we don't have to see the Jews anymore, or how the Germans are doing God's work".' In a recent letter he remarks about Jan Gross's book *Neighbors*, the story of Polish complicity in the murder of hundreds of Jews in one village. 'Gross was right; if anything, he understated the hostility.'

Devastation to body and mind: ghetto isolation

The general public's knowledge about the Holocaust is confined primarily to the death camps. Millions know about Auschwitz, Treblinka, Sobibor, Belzec. This is not the case with the ghettos; place-names like Vilna, Kovno, Bialystok, Mir are unknown, much less the history of the partisan movement in the forests of Lithuania and Byelorussia. The so-called 'passivity' of the Jews in Auschwitz, Treblinka and Maidanek needs to be understood from the perspective of ghettoization and the devastation to body and mind inflicted by unrestrained violence.

Naomi's mother (a nurse at a hospital in a small town near Lodz where the Germans were rounding up patients and children for transport) hid her in the hospital mortuary, in a room bordering a street where Jews were being loaded in a waiting truck. Naomi could hear the screams and the gunfire; but she also heard her sister scream: 'Mother, help me, you saved Naomi, why don't you save me? Mother, you don't love me ...'[31] The next thing Naomi heard was a gunshot and then her sister fell silent. She waited several hours; and during this time German guards threw corpses into the room where she was hiding. Then she saw the body of her sister, tossed onto a pile of corpses. When her mother came to tell her that the Germans had left, she found her dead child lying next to an unconscious Naomi. It was not until they got outside into the light that she noticed her daughter's hair had turned gray. Eventually Naomi and her mother were transported to the Lodz ghetto; within a few weeks Naomi's mother had died. No record is left of her daughter.

Children in the ghetto played games like 'deportation,' 'Führer,' 'Gestapo,' 'shoot the Jew,' and so on. They beg and come home to find that their parents have been victims of the day's roundup and deportations. Husbands and wives are powerless to save each other. After he witnesses his wife being picked up during a roundup, Abraham Lewin records in his diary, 'I have no words to describe my desolation. I ought to go after her, to die. But I have no strength to take such a step ... total chaos ... terror and blackness.'[32]

Weakening of the body, weakening of the spirit, happened simultaneously. Calorie intake in Warsaw, for example, was around 184 a day. Between January 1941 and July 1942, nearly 61,000 people died from malnutrition. Eighty percent of the food entering the ghetto came from smuggling. German soldiers look at this devastated humanity and witness the embodiment of their racial theories. In the town of Komarno, Sergeant Gaststeiger of the 67th mountain rifle section sees 'a city of Jews [whose denizens were] similar to the creatures often pictured in *Der Stürmer*.' Another German soldier describes Poland as the 'land of the Jews, in which whoever travels will be visited by lice... a land, this Poland, that any pioneer will remember, stating, "Ah Poland, it reeks."' Corporal Mathias Strehn notes that on pushing deeper into Poland, 'the stench of the Jews and their beastliness became oppressive.'[33]

Germans saw what they wanted to see; however, ghetto conditions reinforced racial stereotypes. Refugees in the ghettos, whether transported by the Germans or fleeing from action in the countryside, rarely survived more than a few weeks. In the first year and a half of ghettoization, refugees suffered a higher mortality rate than Jews indigenous to the areas surrounding the ghettos. Wealthy German Jews transported to the Lodz ghetto found themselves reduced to begging within three months of their arrival. Away from home, unfamiliar with the new surroundings, refugees suffered not only the brutalization of the Germans, but the indifference of the ghetto itself. Refugees received the worst jobs; the *Judenrate* discriminated against refugee Jews. Most had no family in the ghetto. Often bands of Jews wandering from village to village would find their ways into the ghettos to escape the beatings, killings, extortion and forced labor.

Early in the occupation of Poland, German commanders had complete authority over what to do with Jews. Nothing prevented the commanders from killing them according to personal whim or giving soldiers the authority to kill. No country offered Jews asylum (although Jews could find some safety in the Soviet Union). A few were admitted to Switzerland, but many more were refused by the Swiss government; those managing the difficult escape to Spain were not returned. Roads were heavily patrolled; help could not be expected from the peasants. *Judenrate* were overburdened with caring for the Jews indigenous to their own territories. Villages that had existed for hundreds of years were overrun in a single day.

Yitzhok Rudashevski, a fifteen-year-old boy, describes in his diary the deteriorating psychological conditions within the Vilna ghetto. It is a vivid account of the sapping of will and the destruction of spirit. Referring to the yellow work permits the *Judenrate* distributed, which assured a greater chance of survival and larger food rations, he writes: 'The people, helpless creatures, stagger around in little streets. Like animals sensing the storm, everyone is looking for a place to hide, to save his life.' People lie in the streets 'like rags in the dirt'; how easy, he observes, it is to break the human spirit: 'I think: into what kind of helpless, broken creature can man be transformed? I am at my wits' end. I begin to feel very nauseated.'[34] Referring to the struggle for life, Rudashevski despairs of ever again seeing real human empathy. 'To save one's own life at any price, even at the price of our brothers who are leaving us. To save one's own life and not attempt to defend it ... The point of view of our dying passively like sheep ... our tragic fragmentation, our helplessness.'[35] Ponary, the execution site outside the ghetto, is 'soaked in Jewish blood,' but the 'mass ... goes blindly.'[36]

He refers to the 'blind mass of Jews';[37] everyone is exhausted. 'From the hideouts people emerge like corpses, pale, dirty, with black rings under their eyes.'[38] There are recurring images of little children stealing, their fingers turning blue in the cold, families fighting for scraps of bread. 'Brother was forced to beat brother and to take away from him the morsel of bread which he brings weary with toil to his family.'[39] Theology brings no relief; but religious custom does:

> 'I am as far from religion now [on the eve of Yom Kippur] as before the ghetto. Nevertheless, this holiday drenched in blood and sorrow which is solemnized in the ghetto, now penetrates my heart... . The hearts which have turned to stone in the grip of

ghetto woes did not have time to weep their fill ... poured out all their bitterness.'[40]

The ghetto demonstrates not heroism but sickness, death and numbness. Child vendors: 'frozen, carrying the little stands on their backs, they push toward the tiny corner that is lit up. They stand there [for hours] ... and then they disappear with their trays into the black little ghetto streets.'[41] But they return: and the 'next day you see them again at the sad light, how they knock one foot against the other and breathe into their frozen hands ... frost-bitten blue little fingers ... hands tremble, her whole little body shakes ... ragged urchins with burning little eyes.'[42]

When workshops are closed, creating 'panting and desperate workers ... disheveled and distracted,'[43] or people laid off, panic grips the ghetto. 'People run, people beg not to be discharged; they try using "pull", they cheat, they intrigue. And this commotion is carried over into the house. You keep hearing only about layoffs and certificates and the same thing all over again. People have lost the knack of thinking about anything else.'[44] The loss of livelihood and the possibility of starvation ravages family life in the ghetto, 'a large swamp in which we lose our days and selves,' and pulls people away from engagement.

Dr. Szwajger describes the consequences of this continual assault on the self. 'Deep in our souls, we were all changing, not as children grow to youth with its awakening dreams of romance, nor as men and women pass from the busy activity of maturity to the calm wisdom of age.' Individuals were forced to suffer 'too many shocks, too many horrors, too many changes of the kind that age one even in youth.'[45] The self withers: 'After a while one can no longer weep, no longer love, no longer grieve. Sensibility is numbed, emotions dry up.'[46]

A little boy, disheveled, 'wearing a large pair of shoes on his thin feet ... dawdling and talking loudly to himself' catches a little girl's attention. He is playing a game in which in one hand he clutches a bunch of small stones; with his other he scratches his head. He rushes after Ettie and tells her about the stones, 'nine brothers like these stones we were once, all close together. Then came the first deportation and three of the brothers didn't return, two were shot at the barbed wire fences, and three died of hunger. Can you guess how many brother-stones are still left in my hand?'[47] Ettie, terrified, ran away; but the boy, brought up in a universe and vocabulary of deportation, coupons, ghettos, shots, hunger, workshops, found nothing unusual in this presentation.

During an action in Kovno, in which they entered the ghetto in buses with white-washed windows, full of soldiers, the Germans played nursery rhymes and offered candy to lure youngsters out of their hiding places. Rabbi Oshry recalls:

> 'Mothers who grabbed hold of the bus were driven off by bayonets. Dogs tore at the women's clothing and flesh. One mother, who held on to a bus firmly and refused to be frightened off, was shot through the heart. Her wailing child witnessed his mother's murder. Every bus had the radio inside turned up loud in order to drown out the children's screams. Full buses were driven off, and empty ones replaced them to take on new loads. A number of buses pulled up in front of the ghetto hospital and took away the children there.'

The next morning the Germans returned with bloodhounds and pickaxes to search for children who had been missed during the first action. 'Soon they were smashing walls and cracking floors.' The Germans threw grenades inside anything that resembled a hiding place. The *Kinderaktion* was a commonplace of German policy; 'wild screeching and cries could be heard. And wild laughter, too; a mother had gone insane.'[48] Children had been hidden in cellars, closets, pits, in baskets and bags, pillowcases. 'One of the mothers begged the killers, "Take me along too, I want to go with my child!" The murderers roughly pushed her away and remarked sadistically, "Your turn will come!"'[49]

Terror and fear, the drive for self-preservation, corrupted the ghetto's moral order. The disintegration of moral limits appeared almost daily in the life of the ghetto, with grave consequences for the underground's ability to recruit. In Warsaw the Germans employed Jewish agents to inform about the location of hideouts, the identity of smugglers and black marketeers, the location of valuables. Shop owners sometimes cooperated with the SS or helped in the roundup of those who had no work permits. The Jewish police extorted bribes. The head of the Jewish police, later assassinated by the Jewish Fighting Organization, tore the badges off policemen who tried to save Jews from deportation. In the words of Lewin:

'We live in a prison. We have been degraded to the level of homeless and uncared-for animals. When we look at the swollen, half-naked bodies of Jews lying in the streets, we feel as if we found ourselves at some sub-human level. The half-dead skeletal faces of Jews, especially those of dying little children, frighten us and recall pictures of India or of the isolation-colonies for lepers which we used to see in films. Reality surpasses any fantasy.'[50]

Lewin notes the pervading madness, an insanity that threatens to engulf all life: 'The burden on our souls and on our thoughts has become so heavy, oppressive, that it is almost unbearable. I am keenly aware that if our nightmare does not end soon, then many of us, the more sensitive and empathetic natures, will break down.'[51]

Given this debilitated universe, what hope could the undergrounds or partisans expect from the Jewish ghettos? After the deportations of summer and fall 1942, the Warsaw ghetto had the air of a ghost town, and its inhabitants, specters in a cramped and gloomy universe of dying bodies.

Underground action: self and its restoration

Initially, ghetto undergrounds were a politicized network made up of various ideological movements with continually shifting alliances. Some of the groups in the Warsaw underground included the *Achdut Haavoda*, the socialist-labor section of the Zionist movement; *Akiba*, a Zionist youth organization; *Beitar*, the youth movement of the Zionist-revisionists, followers of Vladimir Jabotinski who advocated militant resistance to the British Mandate in Palestine; the *Bund*, the major Jewish labor union in Poland, centrist party socialist, anti-Zionist and anti-communist; *D'ror*, a Zionist Youth Organization affiliated to *Hechalutz*, the Zionist pioneer youth movement, advocating agricultural settlements in Palestine; *Gordonia*, a Zionist youth group affiliated with *Poalei Zion*, the labor section of the Zionist movement; *Irgun Tzvai Leumi*, the military arm of the Zionist-revisionists (Menachem Begin was a member of Irgun); *Hanoar*

Hatzioni, a Zionist youth group affiliated to *Hechalutz*; *Hashomer Hatzair*, a socialist Zionist youth group with sports clubs and training farms; and *Hitachdut*, a Zionist youth group.

Membership in the political-underground groups consisted primarily of urban, middle- and lower-middle-class Jews, although a sufficient minority of these groups comprised young people who had emigrated from the *shtetl*s (small towns) into the cities. Even with such a heterogeneous group of political-ideological youth groups, these organizations made up a tiny minority of the Jewish population, even amongst the young.

As isolated and desolate as life was in the ghettos, the almost exact opposite prevailed in the undergrounds. Where the ghettos increasingly disintegrated into a Hobbesean panic, with individual families at war with each other, the undergrounds became more and more *dependent* on one another. The more inward-looking and despairing were individuals in the ghettos, the more outward-looking and connected with one another were the underground fighters. The more passive were the refugees, the more active were they when in contact with an underground group. The more the ghetto sank into fatal resignation, the more the undergrounds came to believe in the efficacy of armed revolt. The more death seemed imminent to inhabitants of the ghetto, the more the undergrounds disdained death in favor of the possibility of revenge. As shattering, then, to selfhood as was life in the ghetto, the more resilient selfhood became within the underground organization. Community, in the resistance, sustained life, courage and cooperation – not individualism. Tolerance, open-mindedness, skepticism, suspension of belief, the absence of fanaticism and rage became, in the context of underground life, life-threatening. The more closely integrated were the resistance communities, the greater the probability of survival; the more individuals found themselves advocating action in a secular faith, the more likely were they to seek out and to join a resistance group. Those values we admire in liberal democratic societies – restraint, humanism, reason, limitation, boundaries, the rule of law and respect for the rules of the game – became sources of self- *and* political destruction in the ghettos. The harder the ideological sheathing of the self, the greater the possibility of life, at least in the early days of the underground. The instruction in the use of weapons, the knowledge of smuggling, the self-understanding of one's role as an active resistor, the use of barter with non-Jews to obtain ammunition and food – all these created a self-understanding that began, paradoxically, to restore moral order to the universe. Resistance created an alternative moral environment for the resistors, in which evil did not consume life. As one resistance fighter put it to me: 'the only way I could be a Jew was to kill Germans; the very idea that I could pull a trigger in avenging the death of my family, gave me hope.'

Underground fighters saw themselves engaged in a common struggle. Political organization came to function as moral authority and thus for the participants slowly restored some sense of an ethical and human connectivity with others. Individuals in underground units came to understand themselves not as victims but as avengers in a long tradition of biblical *Jewish* resistance, for example, Masada – which again opened the possibility for a radically different view of self than was common in the ghetto. The diaries of resistance fighters are filled with the imagery of transformation, confrontation and transcendence; the diaries of ghetto inmates reflect a despair even at the possibility of action and a preoccupation with death and moral decay.

Two psychological facts became increasingly clear in what Jewish resistance movements *refused* to accept: first, being drawn into an emotional black hole that would immobilize action; and second, dissociation – a complete affective removal not only from the body of the community but also from the body of the self. Emotional and physical existence in a prolonged numb state, characteristic of ghetto life, never defined resistance attitudes. While a preoccupation with death, which was constantly at the center of ghetto thinking, does not disappear in the underground and partisan units, it is, however, *not* in the foreground of consciousness. What is in the foreground is revenge and to some extent the imagery of rescue, emotions that pull the self together, give it resilience and enable the body to assimilate its suffering and endure the deprivation and uncertainty of resistance life. The group too stands in the foreground: early in underground movements, ideology played a preeminent role in holding the group together, gave the group purpose and constituted the impetus behind group membership. In fighting the Germans one could achieve socialism, further Zionist ideas, train future fighters for Palestine, and so on.

But after the spring of 1942, ideology as a dynamic, a motivating force, falls well into the background. Revenge, anger and rescue, surviving as a group identity, take its place. Ghetto anger, however, often found itself deflected into a bizarre form of gift-giving. So bribes and goods demanded, or even gifts proffered, such as Abraham Tory's offering to the German commander at Kovno a silver cigarette case, were thought to have some effect in sparing (rescuing) life.[52]

The undergrounds and partisans operated from radically different psychological, moral and political assumptions: no gifts, no bribes, no entreaties, no exchanging thousands to save hundreds. It was not only moral to kill, but it was also essential to psychological well-being and to participate in a group whose objective lay in surviving the German plan to kill. What underground and partisan groups understood better than the *Judenrate* was that resistance was not only a route to killing but also a route to rescue of the self and the group, to save Jewish identity from German violence and the intolerable psychological despair induced by German brutality and occupation.

Undergrounds initiated action when Germans advanced on the ghetto; rarely did they undertake actions independently. The Warsaw uprising is the most famous resistance action; but undergrounds also operated in smaller ghettos like Kovno, Vilna, Bialystok and Mir, to name just a few. Undergrounds understood clearly the intent of German policy; they knew what the Germans had in store for them; couriers (primarily young Jews with 'Aryan' features) had brought them first-hand reports of Auschwitz, Treblinka, Ponary and the Ninth Fort (outside Kovno). They understood the German belief system. For example, in Vilna, the underground had publicized the German vocabulary of killing. Murdered victims were called 'figures'; the workers who dug out the graves at Ponary were called 'uncoverers'; those who had to throw iron-hooks into the graves and pull bodies to the surface were called 'figure-pullers'; those who extracted gold fillings from teeth were called 'dentists.' Thirteen-year-old children who picked up scattered bones were called 'bone-gatherers'; those charged with carrying away corpses on stretchers to the burning field were 'carriers'; the prisoners who had to place corpses in a pile, 'burning-masters'; those pouring oil on the corpses were called 'fire-masters'; those constructing pyres, 'pyre builders'; and those who gathered up whatever did not burn, 'gold-diggers.'[53]

Faced with this universe which reduced the Jews to a bureaucratic problem in sanitation management, the Vilna underground pleaded with the populace, exhorting them to fight. An article in the April 14, 1944 edition of the *New York Times*, under the heading 'Poet-partisan from Vilna ghetto says Nazis slew 77,000 of 80,000,' describes how Germans treated resistance leaders:

> 'A partisan detachment was formed in the ghetto in January 1942. It was headed by a 41-year-old Vilna shoemaker named Wittenberg. The Germans announced that if Wittenberg [a communist leader] did not give himself up, all the Jews [in Vilna] would be killed immediately. The leaders met for the last time in a cellar and drank together. Then Wittenberg went to the Germans. The next morning his mutilated body was found at the ghetto gate.'[54]

Wittenberg gave himself up because he received no support from the ghetto inhabitants or from the *Judenrat*, who insisted he turn himself in and who initially aided the Germans in capturing him. One can only admire the courage and inner strength of this man who gave himself up in the full knowledge that he would be tortured and executed. One can only wonder at the blindness of the *Judenrat* in failing to support Wittenberg when it was clear what the Germans had in store for the Jews.

One underground leaflet read:

> 'The enemy in his fear of rebellion, wants to wipe you out first. He is shooting your best sons, taking them to Germany by the hundreds and the thousands. In Warsaw, Kalisch Vilna, Lwow, the Hitler terror has already seized the masses. Do not let yourself be led like sheep to slaughter.... . Thousands of you fall as passive victims. But freedom is bought only by active sacrifice.'[55]

But the underground was powerless to overcome the fear and uncertainty created in the Jewish populace by *Judenrat* caution. For example, in August 1943, the Vilna *Judenrat* warned the ghetto, once again, of the German tactic of collective responsibility and mass reprisal: 'We want to remind the ghetto of the words and warnings of the Ghetto Representative, that as of the latest order received from the German Authorities, all Jews are collectively responsible. It is your duty to yourself and to the ghetto to inform on any activity which might endanger the existence of the ghetto.'[56]

One month after this pronouncement appeared in the official bulletin of the Jewish ghetto administration, the United Partisan Organization (underground in Vilna) distributed the following leaflet:

> 'For our forefathers. For our children who have been murdered! In repayment for Ponary. Hit the murderers! In every street, every yard, every room. In the Ghetto and out of it, Hit the dogs! Jews! We have nothing to lose! We shall save our lives only if we destroy our murderers. Long Live Freedom! Long Live the armed revolt! Death to the murderers!'[57]

The United Partisan Organization fought against tremendous odds. In the German offensive against the ghetto in September 1943, underground units engaged in defensive actions against the Germans and inflicted casualties; and when defense of the ghetto itself became impossible, many of the units fled to the forests and joined partisan groups. Posters plastered on walls throughout the ghetto declared calls to action:

> 'Do not believe the false promises of the murderers. Do not believe the words of traitors. Whoever leaves the ghetto gate has one path – to Ponary. And Ponary is death! Jews, we have nothing to lose, because death will come anyway. And who can still think that he will remain alive while the murderer exterminates us systematically? The hand of the hangman will reach each and every one. Neither experience nor cowardice will save your life! Do not hide in secret places and the *malinas* [hiding places, bunkers]. Your destiny will be to fall like rats at the hands of the murderers. Only armed defense can save our life and honor. Jewish masses! Go out on the streets! Whoever lacks weapons, let him take an ax, and if there is no ax, grab some iron, or a pole or a stick!'[58]

48 *Jewish Resistance during the Holocaust*

But in the ghetto itself, with the exception of a very few fighters, no one listened to such proclamations; and the desolate self could not make the leap from internal dread to effective action. For most ghetto inhabitants, Itzik Wittenberg and what he represented meant suffering, not liberation; and it was impossible to find amongst the Jewish masses sympathy for Wittenberg as expressed in a song popular with the underground:

> 'The night is foreboding, there's death lurking round us ... the ghetto is restless, and Gestapo threatens our Commander-in-Chief. Then Itzik spoke to us. His words were like lightning – "don't take any risks for my sake. Your lives are too precious to give away lightly." And proudly he goes to his death!'

– a very different example than Joseph Gens's, the head of the *Judenrat*, who, in spite of his extensive cooperation with the enemy, was murdered by the Germans when he outlived his usefulness.[59]

Ghetto authority and underground action

In September 1942, Gens, spearheading the ghetto administration, issued a proclamation regarding the kind of punishment (torture, execution, execution of one's family) ghetto inhabitants who tried to escape to the forests should expect from the Germans. It read:

> 'Six Jews ran away from the Bialewaker Concentration Camp. The German command decreed to shoot ten Jews in the same Camp for each runaway; that is, 60 adults (not counting children). The punishment was meted out. Sixty adults and seven children were shot in the above prison camp.'

The example had not been lost on Gens: 'A similar punishment awaits the population of the Vilna ghetto should a similar thing occur here.'[60]

Not to be psychically broken by such 'examples' of discipline required a deep reservoir of faith in action. Sympathizers of the Vilna underground grew in number throughout 1943. The 'average' Jew who joined the underground began to steal – not from each other, but from the Germans and Lithuanians, 'everything from food to military goods, shoes, clothes, paint, buttons, tin or whatever else was handy.'[61] Weapons remained in short supply as it was almost impossible to steal them, and the price for weapons, in whatever condition, was exorbitant. Yet, 'when the hope of being saved from death in the ghetto grew dimmer, when the Jews began to realize that they had nothing to lose, the number of individuals who began to risk their lives by stealing even weapons began to grow.'[62] Yet it remained difficult for the underground to recruit support.

As early as January 1942, the Vilna underground warned the ghetto of German intent: 'Of eighty thousand Jews in the "Jerusalem of Lithuania" [Vilna], only twenty thousand survive. Before our eyes have been torn from us our parents, our brothers and our sisters... . Hitler aims to destroy *all* the Jews of Europe. It is the lot of the Jews of Lithuania to be the first in line.'[63] Leaders of the Vilna underground, Joseph Glazman (Revisionist), Itzik Wittenberg (communist), Abba Kovner (*Hashomer Hatzair*), Abrasha Chwojnik (*Bund*), Nisr Resnik (General Zionist), Major Isidor Frucht (non-aligned) and Chiena Borowski (communist) found themselves blocked, at every step, by *Judenrat* fear and the mass's desolation and emotional isolation.

As early as December 1941, a month before the underground agreed on a course of action, Aba Kovner articulated the mood of the ghetto and attacked the delusional belief that the genocide might stop. The ghetto masses, he argued, 'believed that while they would face a life studded with vicissitude, the slaughter of millions was outside the realm of possibility.'[64] But this belief, which the *Judenrat* fostered, kept news of the executions at Ponary from being a catalyst for revolt. At a meeting of the Pioneer Youth in the ghetto on January 1, 1942 – and this is before the Warsaw deportations to Treblinka and Sobibor – speaker after speaker emphasized that those taken from the ghetto ended up not in a work camp but at the execution site. Yitzhak Arad repeatedly stressed in his analysis of the political reality of Vilna how 'large sections in the ghetto identified with the *Judenrat*'s course of action,' insistent on following German orders and not subverting German policy. In addition, many Jews believed that it was only communists who would be killed by the Germans; and if killing communists preserved the remnant, then it was a cost the populace was willing to bear. Ordinary, non-ideological people would be spared – a delusion desperately clung to even in the face of the facts.

The underground in Vilna set out its goals in January 1942 – to engage in sabotage, to resist if the ghetto were invaded and to establish links with partisan groups and undergrounds in other ghettos. For the underground it was moral and just to collect arms, even though the Germans made it clear that the discovery of weapons would result in mass reprisal. Kovner writes: 'Had we the right to endanger the lives of the thousands of remaining Jews in the event of the discovery of arms in our possession? With full realization of the responsibility we bore, our reply was: Yes. We are entitled, we are bound to do so.'[65] As late as August 1943, Gens published the following – and this after the Warsaw ghetto uprising: 'May the blood that has been spilled be a last warning to us all, that we have but one way – the way of labor.'[66]

Gens's policy of collaboration influenced the ghetto to the point where ghetto fighters found themselves on some occasions betrayed by fellow Jews. For example, in September 1943, a brigade of 100 fighters of the United Partisan Organization was surrounded and killed as a result of treachery by the Jewish police and a Jewish informant. Yet, Arad, active in the underground, paints in my interview with him a somewhat more nuanced moral picture: 'You had to have some sympathy for Gens; what else could he do?'

The UPO leadership in Vilna constantly wrestled with potential traitors and with their moral responsibility to the Jewish masses. Kovner:

> '"As regards revolt, we cogitated more than anything else over the moral aspect." Was revolt legitimate in view of the fact that the majority of the ghetto would not support armed resistance? "Were we entitled to [fight] and when? Were we entitled to offer people up in flames?"'

The underground realized that few Jews had weapons or even a desire to engage in physical resistance. 'Most of them were unarmed – what would happen to all of them?' And what if the fighters were wrong; what if the roundups did not mean 'liquidation'? 'We were terribly perplexed as to what right we had to determine [the mass's] fate.'[67] This moral quandary, always at the heart of negotiations between the UPO and the *Judenrate*, inhibited their ability to act and to plan even with their knowledge of mass executions. Eventually the UPO leadership came to the realization that any hope of mass defense was an impossibility: 'There is no longer any hope that the battle, which a handful of fighters, limited in number, would initiate, could turn into a mass defense... . The rebellion, should it break out, would be nothing but an act of individuals alone, of no wide-national value and would not open the door to mass rescue.'[68] With few exceptions, such as Warsaw, the Jewish population had been effectively silenced by fear of mass retaliation.

The Jewish community and depletion of will

To understand the state of mind of the Jewish ghettos, it is essential not to underestimate the power of the German assault on the Jewish body – beginning long before the construction of gas chambers at Auschwitz. In Warsaw, food distribution early in the occupation assured a slow death. In January 1941, the weekly ration of sugar was: for Aryans 16 oz, Jews 4/5th oz; fruit juice: Aryans only; soap, Aryans only. After the first week in January, Jews received no sugar ration; during the first six months of 1941, Jews received 3 oz of bread daily. Sugar, butter, eggs, fat, vegetables, milk had to be smuggled in and bought on the black market. Even the 3 oz of bread Jews were to receive daily 'means nothing in reality, for we are never able to obtain it.'[69]

Poles were proscribed by the Germans from selling merchandise or food to the Jews; if a Pole were caught engaging in such transactions, the Germans could impose a fine of 1,000 zloty.

In 1941 in Warsaw, 80 Jewish soup kitchens dispensed 120,000 meals every day; the children's soup kitchen run by the Jewish Self-Help organization served 35,000 meals daily to starving children, many of whom had been orphaned. The soup kitchen meal consisted of a thin soup with a small piece of bread. The bread administered by the Germans contained 33 percent sawdust. By 1942, the number of soup kitchens had increased to 145, and children's kitchens to 45; Jewish Self-Help dispensed 60,000 bowls of soup to the elderly, the sick and children unable to walk to the soup kitchen.[70]

The *London News Chronicle* of May 1942 published the account of a woman who had fled Warsaw and escaped to Palestine.[71] It was her estimate that 10,000 Jews were dying every month.[72]

Between February and April 1941, over 44,000 refugees arrived in Warsaw; of the thousands of children in shelters, more than 42 percent were infested with lice. In the summer of 1941, the health of the refugee children was deplorable: just 13 percent were in good physical condition, 35 percent were 'tolerably healthy,' and 52 percent were in poor to bad health. By early 1942, their health status had worsened considerably; now only 30 percent were in good health and 65 percent were in very poor health; 54 percent of the children examined were 'filthy and full of lice.'[73] By late 1941, the death rate amongst all children in the ghetto was between 25 percent and 35 percent, depending on whether the children were in hospital, shelters or living with parents or relatives in rooms or cellars, and with a great many on the streets. A year later, more than 95 percent of all children in the ghetto had been murdered – either through starvation, disease or extermination in the death camps.

To give some sense of the magnitude of these death rates, compare the following: in 1941 the yearly death rate of patients in hospitals in the USA was 3.9 percent. In the Jewish Central Hospital in Warsaw it was 20.3 percent; in the Jewish children's hospital, 24 percent. In 1941, 47,428 Jews died in Warsaw.[74]

What was required was a political vision, and there was plenty of that in the partisan units in the forests and in resistance organizations inside the ghettos. Indeed, it was the failure of political vision, and the reliance on historical practices of accommodation, that contributed to the doom of the Jewish population of Europe. Given the acquiescent character of the historical and religious traditions of 'resistance' in dealing with the enemy, the political vision of the secularist found itself swamped by the hopelessness of starvation and death and the annihilation of children.

It is, of course, true that massive resistance, given the circumstances, would have been extremely difficult; the illusion of saving the remnant which the Germans fostered, the historical tradition of non-violence in the Diaspora community, the lack of any significant assistance from the surrounding Polish population, made resistance appear to be hopeless. Ringelblum persistently heard criticism of Jewish inaction from his contacts with the Polish resistance; yet, the

Polish resistance had no first-hand contact with the ghetto starvation, demoralization and dislocation following resettlement. To have expected anything like mass resistance was itself an illusion.

Yet, other political and psychological factors need to be accounted for in the politics of ghetto administration. The *Judenrat* discouraged the Zionists and communists from organizing and gave little money for active self-defense; indeed, well into the massive deportations in spring and summer of 1942 in Warsaw, the *Judenrat* operated on the remnant mentality. It would be wrong, however, to argue that inaction derived only from causes within the ghetto. The German war on Jewish children and infants, the vast relocation of rural populations to urban ghettos, the despair caused by daily, random deaths, corpses lying unburied in the streets, the insidious German manipulation of the *Judenrat*, the killing and brutalizing of the rabbis, contributed enormously to a collective state of mind in which the masses were convinced that no matter what they did, they were doomed.

Those who worked with various resistance organizations within the ghetto managed to purchase some weapons, but they were of inferior quality, very expensive and, by 1942, very difficult to come by. In Warsaw, Ringelblum writes: 'After long, very long efforts, arms were received but in such a small quantity and of such bad quality that there was no possibility of undertaking any [collective] defensive action.'[75]

The situation was made worse by what appeared to be an increase of Polish anti-Semitism during the war. A Polish resistance fighter, Aurelia Wylezynska, active in aiding Jews, wrote in her diary: 'A wave of anti-Semitism has engulfed the Polish people ... . We are surrounded by a nest of vipers, characters from the underworld of crime... . For every hundred evil men, it is hard to find even one noble soul.' Another Polish resistance fighter, Adam Polewka, wrote shortly after the war: 'The Germans will throw stones at Hitler dead, because he brought about the downfall of the German people, but the Poles will bring flowers to his grave as a token of gratitude for his freeing Poland from the Jews.' With sentiments like this amongst the Polish population, the necessary relations between a broad-based supportive popular movement, willing to give aid to a Jewish mass resistance, could not have existed.[76] It should also be noted that many individual Poles gave assistance to Jews;

Ringelblum's diaries, for example, refer time and again to courageous Poles who offered assistance, support and shelter from the Germans. But he also, and as frequently, expressed bitterness at the indifference, or worse, the support of the vast majority of the population for the Germans' war against the Jews.

Inevitably, Ringelblum's scorn returns to the *Judenrat*:

> 'The fairy tale about the "resettlement" in the East supported by the *Judenrat* and by the band of Gestapo agents brought in from Lublin [informers supplying the Germans with information and spreading false rumors], was so widely accepted by the Jews that thousands of people who were starving as a result of the constant cordons and the complete stoppage of smuggling presented themselves at the *Umschlagplatz* [the train depot in the ghetto transporting Jews to Treblinka and Sobibor] *voluntarily* in order to be sent to work in the east.'[77]

3
The Moral Position of Violence: Bielski Survivors

By the beginning of 1944, the Bielski Brigade consisted of 1,200 survivors. It was led by Asael, Tuvia and Zush Bielski, gathered in a family camp, in the middle of a dense Byelorussian forest and organized as a self-sustaining community. Eight hundred and fifty people functioned in supportive roles, with around 300 or so active resistance fighters armed primarily with rifles and revolvers. Unarmed refugees, mostly women, old people, children, the sick and wounded, lived side-by-side with the armed fighters. The Bielski Brigade demonstrates the success of resistance groups outside the ghetto for whom *rescue* was as important as vengeance. For the two and a half years of its existence, the Bielski partisans lost only fifty members, a phenomenal fact given that mortality rates amongst resistance groups were well over 60 percent. A casualty rate of less than 5 percent can be attributed to the work of the support community and Tuvia Bielski's philosophy of 'rescue,' as well as 'clever politics' in dealing with the Russians and the prowess of the fighters. This *shtetl* in the forest exacted vengeance on the Germans, while encouraging caring and cooperation as the primary objective of a civilized and human existence. Rescue and community for the Bielski group were as important as killing.

In looking at the Bielskis, I am concerned with the moral position of resistance. The history has been adequately covered by Peter Duffy's *The Bielski Brothers* and Nechama Tec's more scholarly, sociological analysis, *Defiance: The Bielski Partisans*. For a complete study of how the Bielskis operated and an account of the structure of their community, both Duffy's and Tec's books are essential reading in understanding partisan resistance. What I focused on in my interviews with Bielski survivors were the moral gray zones, problematic intersections of violence and surviving, what Zvi Bielski called the brothers' 'mayhem,' in addition to the impact of memory, particularly memory involving spiritual and theological states of mind, and how survivors explained their luck in having lived to establish families and build new lives.

The Bielski group initially started with a few relatives and friends who had survived the massacres in Novogrudek, a small village in Byelorussia; by 1943 it had swelled to well over 1,000 partisans. Yet, unlike almost every other partisan group, both Jewish and non-Jewish, the Bielskis insisted on accepting any Jew who sought refuge with them. Tuvia Bielski refused to turn away any refugee – no matter what their age or physical condition. But even more important, all found a role in the group, and these support roles – what one would find in any village or town, from food preparation to laundry, to maintenance and supplies, and tool-making and tannery – rather than draining energy from the fighters, contributed to the community's overall welfare. Cooperation, not desolation, informed the group's psychological environment.

Tuvia Bielski's leadership consistently emphasized the welfare of the group. While conflicts arose over specific issues – such as who kept 'luxury' goods (for example, meat) taken during food expeditions – the group obeyed few fixed rules: all basic foodstuffs had to be handed over to the communal kitchen. Fighters, often led by Zush and Asael, who had been on food-gathering expeditions, might divide up choice commodities (such as a cow) before returning to camp, and support people would complain about this, but no one went hungry. Sonia O.: 'Even if it were only a piece of bread, it was something; a little bread and water went a long way in the forests.' Everyone received at least two meals a day: breakfast, consisting of boiled chicory and potatoes or bread; and a fairly substantial main meal, though the quality varied, depending on conditions and availability. In addition, there were those with access to better quality food: fighters, communal leaders, those in a position to occasionally supplement their allotment by preparing food in their own quarters.

A social hierarchy prevailed in the detachment, with commanders at the top; next came fighters; then people with special skills. At the very bottom were those without any forest survival skills. A young girl describes her father's position at the bottom of the hierarchy, who before the war had been a well-paid administrator of a large brewery: 'In the *otraid* [detachment], he became a "*malbush*" [someone who did not fight, and who offered the group no special skills]; he did nothing ... . He was intelligent, educated, but not at all resourceful.' But non-combatants and those who did not work were never left to die; nor were they left to starve. Nonetheless, a great deal of social opprobrium was attached to them. Max: 'We had little use for the *malbush* ... so many people eating and drinking and having a good time; we fighters resented them.' Max appeared to be much closer to Zush's attitude to rescue than to Tuvia's.

Yet, the story is more complex than that. Zvi Bielski tells me about his Dad's rescue of some *malbush* when the unit had had to hastily retreat deep into the forests in the face of a huge German mobilization in search of the partisans. Zush told one of the group to keep an eye on a woman and her daughter, and to make sure both escaped the encampment. Zush found out that the man had gone with the group but had left both the mother and daughter behind. When he asked him why, the man responded that they were not his concern; why should he try to save them? Enraged, Zush screamed that they deserved to live as much as anyone in the group, pulled out a gun and shot him. Then, at great risk to his own life, he went back, found the mother and daughter and they rejoined the unit.

Malbush could improve their position through various means; for example, by guard duty or working in the tannery. Where one came from, and how much time had been spent in the forest, also affected social standing. The clothes one wore, for example, the fighting unit's dress, and ownership of weapons, distinguished them from others of the *otraid*. But carpenters, bakers, tailors, gunsmiths and tanners, though on a level below the fighters, were highly respected for their skills and contributions to group survival.

In the fall of 1943, a permanent base was established in the Nalibocka forest. Workshops were housed in buildings dug out of the ground, and the base took on the properties of an organized community. Services, provided free of charge, kept the group functioning as a community; and all were entitled to have personal effects, including weapons, clothes and tools, repaired. Transactions involving other partisan groups had to be arranged through the leadership, and these involved payments of some kind: food, supplies, weapons, and so on. One survivor recalls: 'Most workshops were situated in a very large hut. The din emanating from this hut could be heard from afar, banging of sledgehammers, sawing of wood, clatter of sewing machines, laughter, lively conversations rich in partisan slang.'[1] In the ghettos workshops like these helped the Germans; in the forests the workshops benefited the partisans:

> The huge hut, with its raised ceiling which looked like a large machine shop in a factory, accommodated tens of workers, who were divided up according to their trades. Large windows provided proper lighting for the various workshops located in all corners of the hut. The different workshops were separated by wooden partitions, and a number of people worked in each cubicle. More workshops were spread out throughout the camp … . All materials provided by nature in the forest were put to use.[2]

People who had goods to trade were served first; those with nothing had to wait, sometimes for long periods, whether it be for the tailor, the gunsmith, the carpenter, the barber, the tanner. But no matter how long one waited for a service, it was always provided. Guards were constantly posted, yet in an environment of hostility and uncertainty, the Main Street character of this village in the forest lent stability and meaning to a world of despair and death. Nevertheless, many of the fighters saw the support group as a drain on partisan operation.

Many Russian commanders were suspicious of the Bielski group and wanted it disbanded. However, when General Platon, the Russian in charge of the partisan brigades in the Baranowicze region, toured the camp, he was impressed by the model partisan community he found and insisted that its integrity be preserved in the Soviet partisan system. According to Tec: '[Platon's] insistence that the Bielski *otraid*'s contributions were essential to the partisan movement saved the life of Tuvia Bielski.'[3] Politics mattered, but in Zvi Bielski's words, fierceness mattered more: 'The Russians knew the Bielskis were prepared to fight to preserve their unit, to kill Russians if they had to.'

Women were treated on equal terms in the Bielski group, although they rarely were allowed on partisan missions. Much of that reluctance had to do with cultural and traditional attitudes; and while some women had guns and used them with partisan units, the majority in the Bielski group, as in other partisan brigades, were assigned to support services. Sonia Bielski: 'I had a gun but Zush didn't want me going out on missions.' Yet, with few exceptions, women refused to allow themselves to be treated as second-class citizens even though most of the fighters resisted efforts to bring more women into the fighting units.

Tuvia's spontaneity, his accessibility to the detachment, his tolerance when necessary, in addition to astute decision-making and his willingness to kill to assure the group's survival, constituted a form of leadership that was absent in the ghettos. To protect his detachment from Russian anti-Semitism, he constantly reminded Soviet officials of their government's policy of nondiscrimination; that Jews were Soviet citizens like everyone else, in addition to being patriots who in the forest were defending the Soviet Union and the unity of its system. Zush, according to his son, in not so subtle ways reminded the Soviets that these fighters would not hesitate to kill for their survival. When Platon visited the Bielskis, he saw some elderly Jews *davening* [praying]. When he asked what they were doing, Tuvia said they were reciting the words of a Soviet patriotic song. Tuvia had a talent for managing strong personalities, an ability that improved relationships between different factions in the brigade and mediated differences, including those with Soviet brigades that on occasion threatened the community's capacity to function.

Tuvia's insistence on rescue rather than revenge as the community's fundamental rule was the dynamic primarily responsible for its survival. In the words of one survivor: 'It seems to have been the right decision. Had we remained a small group we could not have made it. So the goal was to become a big group.' While many of the early survivors, including Zush, opposed this decision, 'Tuvia insisted that if we were bigger we would have a better chance of survival and we would be more secure.'[4] Rescue, then, took precedence over killing, a fact that distinguished Tuvia's detachment from Jewish and non-Jewish partisan brigades built *only* on the pursuit of vengeance and who refused admittance to anyone without a weapon and the physical stamina to withstand partisan life. This is not to say that Tuvia avoided violent missions. They were as essential to his group as to the other partisan brigades; and if he believed he had to fight, he never refused to engage in combat. Yet, Tuvia, in spite of differences with his brothers, saw the community's ability to sustain fighting and to remain intact as a *Jewish* brigade to be tied up with its role as a place of rescue and refuge.

Abraham Viner, a partisan who worked with Tuvia Bielski, remembers him as someone who 'devoted his soul, his brains and everything else to the rescue of Jews. He saw a chance, a great opportunity, in his ability to save.'[5] Another survivor/partisan, echoing sentiments expressed in my interviews with Bielski survivors, writes:

> 'For forty years, we had discussions about what was more important, fighting the Germans or saving Jews. We came to the conclusion that our heroism was not heroism. When I was fighting with guns together with other partisans, this was not heroism. Heroism was to save a child, a woman, a human being. To keep Jews in the forest for two years and save them, this was heroism.'[6]

Yet, divisions remained over the years. Elsie S.: 'I remember at a Bielski survivor gathering several years ago ... there were a number of people there... . One group sat off to the side, refusing to mingle with the others. I asked them what was wrong; why they were not talking. One of them said to me, "Oh, they're *malbush*; we don't have anything to do with them." ... This after 30 years!' One survivor remembers how her four-year-old son always used to ask: 'Mommy, under which tree will be our house today?'[7] But the trees brought safety. Elsie S.: 'I loved those trees; and the trees told me stories; I spent hours speaking with them.'

Conflicts over leadership, periodic disputes, required Tuvia's intervention. But even with sometimes serious arguments, the group survived and grew. The fighting brigade fluctuated between 20 and 30 percent for the entire detachment. Most of the brigade came from lower-class backgrounds with only a small minority from the upper or middle classes. Most had little or no formal education since few of the Jewish elite survived the initial German occupation and mass executions; and the few who did survive and managed to escape were ill equipped for forest life. Those who prior to the German invasion had been considered 'lower class' now became the elite. According to Tec, 'Physical strength, an ability to adjust to the outdoors, and fearlessness were qualities that mattered. A man's prestige depended on the extent to which he exhibited these qualities. Women were usually not included in these calculations,'[8] although lower-class men sought out upper-class women as forest 'companions' or 'wives.' In Sonia Bielski's words, 'We were young; we didn't know if we were going to be alive tomorrow. So, love came to us quickly in the forests; we needed some happiness.' Working-class people adjusted better to the physical stress of the forests, and looked with contempt on educated Jews who struggled with the rigors of forest life. Aaron (Bielski) Bell: 'Look we grew up in the woods; it was our natural place to live. City-Jews had a rough time with this.'

A refugee first encountering the Bielski camp recalls:

> 'I was amazed ... I thought that it was all a dream. I could not get over it... . there were children, old people, and so many Jews. When the guard stopped me, I spoke Yiddish. I met people who knew me. That first time I could stay only an hour. After a few days, I went back and then again and again... . Once I saw a roll call, soldiers stood in rows, with guns. I saw two men come out, tall, handsome, leather coats... . I asked who they were and was told that these were the Bielski brothers. They were giving orders to the fighting men. They were going for an expedition, Tuvia and Asael ... The two jumped on the horses like acrobats. I imagined Bar Kochva to look like that ... Judas Maccabe, King David... . It gave me hope.'[9]

While medicine was in short supply and a crude hospital took care of the very ill, death from illness and exposure was almost nonexistent. Infections were treated with injections of boiled milk and, if available, iodine; the sick received extra food rations, and the group's only typhus epidemic claimed just one life. People managed to stay healthy; the major physical maladies the group faced consisted of skin diseases – blisters, scabies, boils, and fungus infections. Because of its sulfur content, gunpowder was used as a disinfectant to treat infections. Always short of food, medicine, ammunition and weapons, the Bielski group developed ingenious methods of extortion, theft and smuggling, so as not to run foul of Soviet rules regarding expropriation from local populations. Yet the relationship of the Bielski brigade with both Soviets and locals was always tenuous.

The Bielskis' success at rescue and survival remains a powerful story in the history of Jewish resistance; and this brief synopsis is not meant to be exhaustive. The micro-history of the brigade is told as a straightforward narrative in Peter Duffy's *The Bielski Brothers*. This is a terrific account of this remarkable group; but Duffy rarely moves outside the historical story itself. In my interviews with Bielski survivors troubling ethical issues arose, ones that distinguish the made-on-the-spot ethics of the resistance groups from traditional moral constraints operating in the ghettos, which, if maintained in the forests, would surely have meant death. What the Bielski survivors argue is that radical reversals of ethics were absolutely essential to sustain life; yet while never directly questioning the Bielskis' methods, some survivors remain troubled to this day by what they experienced and witnessed. Yet, they also are quick to point out that to have held on to a morality of humanism or hope that others might rescue them in the forests would have been suicidal. It is against the backdrop of this moral drama suggested initially to me by Zvi Bielski's account of his father, that I conducted my interviews in south Florida with a group of brigade survivors.

The Bielski survivors: the past in the present

It was hot in south Florida. The temperature broke records, a muggy moist heat enveloping you and not letting go. Only the air conditioning dispelled it; but how long can you stay in chilled air after the coldest winter in the North in over a century? The heat felt good, but confusing, too much glare, too bright. I felt out of synch and had no idea how to navigate this place: tight, narrow roads, stop lights at every corner and a sky so blue it hurt the eyes. I was determined to find Sonia Bielski. I had directions to her apartment, south on 95, but Mrs. Bielski had given me the wrong exit, the old exit. I call her; she keeps saying 'exit 1,' but exit 1 puts me on a road heading straight for a light industrial area.

My interview with Sonia Bielski, Zush's wife of 53 years, had been set for the weekend after the worst storm ever in Baltimore, two and a half feet of snow, four days shut in. So, I cancel my plane reservations, call Mrs. Bielski, and reschedule the interview for two weeks hence. Three days before leaving, I call to confirm. 'But I thought you were coming last week; I waited the whole weekend for you.' I call her son, frantic; she had misunderstood me. But Zvi assures me: 'She made a mistake ... Just tell her when you will be down.' I call again, apologize and speak of the difficulty of scheduling times. 'Don't worry ...,' she tries to make me feel better: 'Just have a good appetite when you get here.' Even on the phone, it's hard not to be drawn to Sonia Bielski, a strong, determined voice, impatient and quick. She manages to scold and comfort at the same time.

It is with considerable anticipation that I look forward to my interview; the hotel clerk gives me directions to Interstate 95, and I'm off. Earlier, Mrs. Bielski had given me directions to her 'house,' so I assume she lives in a townhouse or a small single-family home. But I get lost; exit 1 takes me to the middle of nowhere, and after several convenience-store stops for directions, I find one of the thoroughfares near where she lives. I look up and what faces me are thirty-story condominiums, three or four of them reaching into the sky. I'm thinking 'house' and all I see is an endless array of huge white buildings. I feel ridiculous; here's a woman who fought with one of the greatest heroes of the Jewish resistance, and I can't even find where she lives.

The street curves in a huge horseshoe, and I drive back and forth looking for 1761, the number she gave me. But the numbers on the buildings read 7611, 7633, 8215, and so on. I reach for the cell phone: 'I'm sorry, Mrs. Bielski; I think I'm lost.' 'Where are you!' I try to describe the scene; 'You're near it,' she says. 'Just find 1761. It's not rocket science!' She hangs up; but 1761 never appears; the numbers, each etched on stone slabs in front of the buildings, leap higher, not lower. So feeling quite helpless and dumb, I again call: 'Mrs. Bielski, I'm sorry, but I'm still lost; I just can't find your house.' 'Professor Glass, I thought you were a smart man; it's right there, on the left.' 'But Mrs. Bielski, all I see are larger numbers like 7632, 7611.' She pauses for a moment; my car stalls in the middle of the road; people sound their horns; I'm flustered. 'Oh my, I'm sorry, I transposed the numbers; my house is at 7611.' I pull over to the side of the road. 'Thank you, Mrs. Bielski, I'll find it; but one more thing; you said a house; do you live in a high-rise?' 'Yes, yes, now you find it. Bye.' And she hangs up, exasperated, I imagine, with this clumsy college professor who seems to have great difficulty finding his way.

I felt I had blown the interview; but, there in front of me, finally, stands her building, 7611. I drive up the huge, expansive stone driveway impressed by the Rolls parked in front; a valet takes the car; the doorman lets me in. And before me stretches an elegant hallway, stately rooms with plush furniture, marble floor, gleaming surfaces, quite a distance from Novogrudek and the violence of resistance. The desk clerk rings Mrs. Bielski, who instructs her to send me up.

A short, energetic woman of 80 with a strong Yiddish accent greets me at the door. She motions me in and the first thing she says is: 'I want you to meet the family.' She takes me to the bedroom and I half-expect a roomful of people. But she points to the dresser full of photographs; her husband Zush as a young man in the forests, and several of him later, in Brooklyn; her sons and their families. She tells me how proud she is of all their accomplishments and the talents of her grandchildren; it is a handsome family with this loving mother and grandmother smack in the middle of many of the pictures. Her eyes shine as she looks at this dresser full of history; and in the effusiveness of her words describing each scene, she brings the snapshots and portraits to life. Her words, in rapid-fire cadence, possess physical properties, heavy, tangible. She composes rich images which envelop you like a holograph; if you raise your hand, you might catch a cluster of words. I'm transfixed by this language, drawn into it; her words surround, confuse and transport me to places in these snapshots, especially those from the forests. It's like a spell. Then I hear, 'But you must be hungry.'

Zvi, her son, had warned me about this: 'She'll feed you and then feed you till you drop.' She takes me to a small kitchen and asks again, 'Are you hungry?' but the question translated means, 'You must be hungry.' I say, 'No thank you, Mrs. Bielski, I just ate breakfast.' My words fly by her, lost somewhere in the wall. 'Of course you're hungry; what do you want, a good "Florida" bagel? Coffee maybe?' She takes out a bagel and I say, 'Coffee is fine, Mrs. Bielski.' 'No, you eat this bagel, and I have some good low-fat cheese, a peach here, nice fruit, strawberries. Eat!' like a general giving a command. I'm seven years old again; in my grandmother's kitchen; anyone, no matter how old, would be reduced to seven years old in Mrs. Bielski's kitchen. There's no way to avoid the food. It was a river I had to cross before she would talk. I knew, then and there, that if I refused to eat, she would kick me out.

So here we are: after all this time arranging the interview, I sit at a tiny table, facing a woman who endured hell and survived, who defined her life in terms of decision and action, unable to reach my pen and notebook, blocked by phalanges of cheese, lox, bagels, cream cheese, fruit, and very curious as to what this wonderful person was all about.

Sonia Bielski thrives on action and movement; I cannot, even in those first few moments, imagine her quiet, still. Even when she sits, her persona is like a whirlwind; her eyes, cheeks, brows, forehead, the surface of her skin, move in exquisite, unexpected directions. After ten minutes of conversation, her energy erases her age; and the more we talk, the younger she becomes. Ever impatient with her own lapses in memory and sequence, she seems at times to be scolding herself: 'Memory, wake up, don't make mistakes, get it right!' It would be difficult if not impossible to persuade Sonia Bielski of a point of view she found wrong. And while she never clamped down on a question or perspective, she had no hesitation in correcting or in instructing in the right point of view, although always with politeness or in Yiddish. And since I told her my grandmother and mother never taught me Yiddish, the phrases flew around the table with increasing frequency, especially when her two friends joined the interview, women who had been with the Bielski Brigade.

Being Zush's wife meant that Sonia Bielski had a privileged position in the brigade, for Zush led the fighters. Sonia too carried a gun, but she never accompanied the fighters on missions that involved killing or executions. About the fighters' 'wives': 'Everybody wanted to be queen,' but competition and friction never endangered the unit. When under attack the brigade reflected a unity of purpose and suffered equally. 'There were times when I didn't live like a queen; when we had to run to escape German sweeps. Once we lived in a ravine for days; I lived on raw horsemeat; but we hoped we would stay alive.' I ask her what she meant 'living like a queen': 'I don't really mean that I lived in the lap of luxury; we lived in the forests and tried to survive; but Zush would take care of me. I had a terrible skin infection, and he took me to a peasant's house to recover. If it weren't for Zush and the brigade's reputation, that peasant might have turned me over to the Germans. But he didn't; I stayed there for several days and when I recovered, Zush came and took me back to the Unit.'

For the most part, Sonia's account parallels that of Nechama Tec in *Defiance*; she refers to the drinking, the occasional infidelities, but dismisses them as minor distractions in a brigade concerned primarily with rescue and survival. What struck me about Sonia Bielski was her being, her persona and the strength of her character, a tenacity surrounding this 80-year-old woman. I tried to imagine what this woman was like 50, 60 years ago; what a powerful presence she must have been amidst these determined fighters. When she first joined the brigade, Zush insisted that she be his 'woman' or 'wife'; Sonia, as she put it, 'found him attractive; this tall, handsome man'; but she had one condition. 'I said to him he must rescue my parents from the ghetto before he could have me.' Zush never hesitated; at great risk to himself, he successfully managed the rescue operation.

It is very touching, sweet, this love story between Zush and Sonia; the love, passion, runs through her every word; intensity, commitment, admiration, frustration, jump from her eyes, and the sadness, too, of his no longer being there. But it was a passionate and desperate time; men and women loved each other. No one knew if they would be alive the next morning. Sonia's remembrances were more about love and working to patch over differences than about killing and revenge. 'Yes, at our "wedding" we had 1,200 guests; God gave us the *huppah*.' If a fighter wanted a wife, it had to be now. Leah J.: 'A few days after I met him, my husband told me "you have to be my wife or I will kill you and myself".' But 'luckily, I liked him, so I went with him.' Or, as Elsie S. describes it, 'A girl living with a man had a better chance of living.' Pregnancies generally were dealt with by abortion; but one or two births occurred in the brigade.

About two hours into the interview, Mrs. Bielski invites her two friends over for lunch and the four of us sit round her small kitchen table. It is an extraordinary conversation, moving alternately from English to Yiddish, with the women disagreeing over interpretation and facts, but intent on describing what they endured. I find myself traveling back in time and memory with these women, whose voice and eyes reflect pride and sorrow, anger and wonder at their survival. And in the images they use to describe survival, I sense something like a prayer of deliverance, an affirmation understood through the continuity of generations, the accomplishments of children and grandchildren. Survival in their accounts could not be disconnected from birth, natality. Behind their eyes, and in the sonorousness of their heavily accented voices, and their own wonderment and even confusion at the reasons for their survival, each spoke of the experience in terms of natality, what they subsequently created as mothers and grandmothers. It was a powerful theme that tied past and present together, that allowed them to locate and define the reasons behind their survival.

I had not heard the importance of natality expressed in this way in the narrative of the men, who while speaking of their great pride in their children and grandchildren, never placed them at the moral epicenter of survival. Natality in the narratives of the women stood at the center not only of who they are now, but what they had been 'then.' It was as if they were saying back 'then' they held inside themselves, physically and spiritually, the key to the future; that their bodies could, if they survived, assure, if not sanctify, the continuity of the community through the literalness of birth.

Natality and being Jewish: these themes encircled each other, as if survival had been God's way of saying to time and history, 'It will all begin again, and it begins with birth, with your bodies.' Sonia O.: 'And that's why I survived.' The children represent continuity and connection with a community stretching back into the past and forward into the future, a community greater than themselves, but one that would not exist without natality, the very physicality of their bodies as guarantors of identity. This belief, commitment or faith in natality, creates a biological link to their own murdered families; it is their testament, or *Kaddish*, to the death of mothers, fathers, grandmothers, grandfathers, brothers, sisters, sons and daughters. Natality establishes a spiritual conversation of the living with the dead; because through their children, the remembrance of those they loved moves across the boundaries of time. With birth comes a re-creation of the past; the child brings back into the world the spirit of those who perished. The mother sees in her child an image of her dead mother and father.

The forests bring birth; and it is the omnipresence of birth, spiritual rebirth, not death, that hangs over the forest. For Jews traumatized by German barbarity, the ghetto held death and terror. But the forests allowed for the creation of action, the establishment of a community; it was the space of natality and regeneration. Being a partisan transformed all moral perspective; as one survivor put it: 'No one in the forest was bad; we were all good.' Concepts of good and goodness change in the forests; anti-Semitism, the vicissitudes of survival, alter the traditional formulations. 'Good' consists in surviving, and that means fighting, subverting the aims of the anti-Semites, intimidating peasants and stealing food. 'Bad' defines itself as what anti-Semites can and may do to you. The unrelenting hatred of Jewish partisans by Germans, peasants, Poles, some Russian partisans, forced the Jewish brigades to stretch and redefine the meaning of good and bad. For example, forcibly confiscating food from peasants had nothing do with 'Thou shalt not steal'; it was rather, 'We are collecting taxes.' As Charles Bledzow, a member of the Bielski Unit, described it to me: 'We would come to a farmer by surprise; they know what to do; that was the power of our reputation in the forests.' And then he adds, with a smile: 'We justified it by arguing it was taxes; they paid taxes to us.'

Ethics and morals in the forests possessed meaning only as they impacted surviving; moral difficulties in terms of relationships within the brigade (for example, over food and labor distribution), if they arose, could be negotiated, although internal conflicts on a few occasions reached the point where they exercised a divisive impact on the brigade. Individuals understood the limits, even if they were not explicitly stated. The unspoken dictum meant action threatening the brigade's unity or safety would not be tolerated, whether it came from the inside or the outside. Bledzow: 'You ask about ethics and morals; we had ethics and morals but we invented our own to fit how we had to survive. You had to do things to bring in food; that was it, the bottom line. With many of the Russians anti-Semitic and the local farmers hostile, if you didn't use force, you were a dead man.' Sonia Bielski praised the fighters and non-fighters, but, she said, 'human nature' complicated the picture: 'Everybody was greedy like an animal; we fought with each other; there were conflicts, but we managed to cooperate … . You were bound to have some jealousies … the fighters received the most generous food portions, they had the wives ... sometimes the *malbush* felt frustrated and left out.'

Many of those not part of Zush's brigade but in the main group with Tuvia Bielski fought in their own way by gathering food, guard duty, reconnaissance, and these too involved risks. Sonia O.'s husband, for example, while not a fighter, worked at whatever was asked of him. He had a gun, but never used it; yet she felt safe with him 'because the group made us safe in the forests; we could protect ourselves.'

Sonia O.'s security in the forest involved more than the presence of guns; it also evolved around a continuing faith that God was watching over this community and would save them. 'I knew we would make it; at times I had my doubts; it was no bed of roses, but I prayed I might one day see my children and tell them about their grandparents. Even though so many Jews died, we won by giving another generation.' Sonia smiles; pride fills her eyes. 'But you know, this faith in what my children might bring kept me going.'

Faith in natality played a significant role in maintaining a psychological balance, never succumbing to fear, depression and despair over loss of family. In Leah J.'s words: 'God is in your children; it is a terrible thing to be a Jew... . So the children and the knowledge that you might survive and have children become your future and your salvation.' But there was more to it than that, a negative, troubling counterpoise to natality; the belief that Jews deserved this suffering; guilt in the eyes of God. Sonia O.: 'This was very much on my mind ... maybe God was testing us.' The men I spoke with had little patience with this spiritual position; they rejected it outright. 'Germans killed us; not God.' But, with the exception of Sonia Bielski, the women embraced this concept. Sonia O.: 'Who are we to know God's will? When I raised this in the brigade, they shut me up, particularly the fighters if they were around. So, I thought to myself, maybe we went away from God; we moved from his path.' The sadness in her voice wraps around her words and slows them down. 'Who knows ... it was so long ago ... we did something wrong that God didn't like. And when we survived, it was God telling us to be observant as Jews, to instill faith in our children.'

To prove her worthiness in having survived, Sonia O. argued, she felt she had to demonstrate by faith in God; that meant strict observance, keeping kosher and giving her children religious education.

'There is a God, I know because I'm alive'; yet, surviving meant overcoming the guilt at having survived. 'Why me? Why not my sister, brother, father, son?' Guilt recurs throughout Sonia O.'s narrative, a heavy obligation that interrupts the pleasures of natality.

> 'My spirit sometimes hurts; it hurt then when we had a few moments to think to ourselves. You feel guilty for the crime you did not commit. I didn't kill Jews; but I still felt guilty. It bothered me in the woods; and it bothers me now. Why am I alive? Why were my brothers and sisters killed? Why is my little sister dead and not me?' She is quiet for a moment. 'Maybe I shouldn't have survived; was it luck? ... God's will?'

These thoughts, recriminations, preoccupy her, a remembrance that possesses no answer, no resolution, an open wound in her soul:

> 'The older I become, the more this haunts me. I thought about it once in a while in the forests and later, after the war, there were times when it entered my mind... . But now, this guilt at having survived comes to me all the time. I wake up in the middle of the night, wondering, why me? Why did I survive?'

Sonia Bielski, however, dismisses guilt. 'You have to be a Jew in your own way'; decision, action, healed her soul. 'Being a partisan was my prayer.' She drew her spiritual inspiration from the sight of Zush and the fighters, and the knowledge that they were protecting her. 'Killing Germans: that was our prayer. We knew we had a place; it was in that forest, where we lived; the brigade gave us our freedom; it allowed us to make our own choices; it freed our spirit.' And even though the forests could be difficult, filled with obstacles, 'It was ours; those bunkers in the ground; they were our place.' Freedom and spiritual liberation lay in the protection of the trees: 'We didn't have time to think about what we've lost, only what we could save by fighting. In those forests we were free.' But even she, at moments, had her doubts:

> 'I once asked my husband, "Why is God killing us?" He would have none of it. He told me not to think about it. "*We* kill Germans, collaborators; God has nothing to do with it." But I

thought to myself, God is giving and taking; you have to accept that fact, and then do what you can to fight back. I let my hate of the Germans take over; then I stopped thinking about God.'

Yet, Sonia O. reminded me that the group with Tuvia, the non-fighters, possessed plenty of hate and desire for revenge. The ethics of community shifted after the German assault on Novogrudek. What would have been unthinkable in the *shtetl* took on a spiritual and redemptive valence in the forests; normally mild-mannered and peaceful persons acted violently and with rage.

It is late in the afternoon the next day. I had called Sonia O. and asked her if we could continue the discussion. We are alone in her apartment and she seems eager to talk. Yet her mood takes us into deep shadows. 'Lots of things I don't want to remember ... so much hate on both sides.' Tears stream down her face, but she insists we explore these painful memories.

> 'In the ghetto I was part of a small group determined to find a way out; we had been digging a tunnel. A kid dug with us, but we found out he had been hired by the Germans and police to inform. We later learned he had been responsible for the deaths and capture of several families. Apparently the Germans and police paid him or told him he and his family wouldn't be killed if he told them where to find Jews.'

Her speech slows; she seems to be catching her breath:

> 'There was a barber in our group, a gentle guy; we knew him as a person who never hurt anyone. All he had done in his life was cut hair; but his family had been murdered. One night, after digging, he grabbed an ax and without a word hacked off the head of that kid. I remember his eyes; how wild they were. We buried the head and body in the tunnel. No one ever spoke of it.'

The sadness in Sonia's eyes is not sympathy for the informer, but for the barber, the memory of his bitterness and the transformation of a gentle soul into a killer, an avenger. She looks at me, and says as much to herself as to me: 'Look, the Germans wanted to torture you, to rip out your soul and burn your body. We couldn't avoid that. We had to protect ourselves.' She turns away for a moment and we both sit quietly; she looks to the side, towards a large window which opens out to the next building, no view, but an empty space punctuated by the bright whiteness of the building's structure. Her voice returns, but much softer. I can barely hear.

> 'I remember one more thing. I have never described this to anyone. It was horrible, but it happened. A couple of times, we caught German soldiers running from the Russian front. We murdered them. Everybody took a stick to kill the German. The one that I saw, they cried, "This is for my mother, my son, my grandmother" ... no mercy. Each person, old men, women got in a blow. When I saw this I started to cry, not for the German. I hated him just as much as anyone else. But the Germans killed my father the same way, beating him to death with sticks ... I couldn't bear to watch it.'

Sonia stops speaking and then continues, insisting: 'It's wrong to say only the fighters were violent; all of us in the camp killed Germans when we had the opportunity. But I couldn't watch this; it brought back the memories of my father's death. I ran away, into the woods, not for the Germans. What those people did to the soldier was right. But I just kept thinking about my father.' For many Jews, faith sustained itself on vengeance.

It is critical not to underestimate the power of this community in the forest to hold and contain selves physically and psychologically battered by the German policy of ghettoization and annihilation. Each participant in the Bielski Brigade had lost if not an entire family, then the majority of their loved ones. Lapsing into madness, a frequent occurrence in the ghettos, especially amongst children, was fought against in the forest, but it took an enormous commitment from everyone to self-defense, and from the community itself to hold and comfort each other. In the words of one survivor, Elsie S., 'How could a human mind still be normal in the face of this? ... I thought I was dying; I lost my entire family in a single day ... I was envious when I saw a bird flying; that bird had freedom.' Ghetto diaries, as conditions become increasingly difficult, describe individuals and families disintegrating emotionally; catatonic children wasting away, fouling themselves against ghetto walls; shrieking women roaming the streets; suicides increasing daily. The forest community, however, provided barriers to this sinking into madness. Elsie S.: 'You talked with everyone in the camp. Any work was helpful; you exchange ideas; try to console one another. People, if they wanted, could be understanding. But you couldn't be morbid all the time. How can you function if you think about death every minute of the day? It was a new way of life.' In the words of the youngest Bielski brother, Aaron, twelve years old at the time of the Brigade's founding: 'With the Bielskis in your camp, you felt you had an army; we knew that to be in the forest, it was far better than the camps. Compared to those living in the concentration camps and ghettos, we lived in the Waldorf Astoria.' Children in the Bielski Brigade fared far better than children in the ghetto, although children never received any formal schooling. Still, the few children in the brigade played with one another, and adults cushioned them from the trauma suffered during the German assaults.

Aaron (Bielski) Bell, eleven when the German invasion began, witnessed the crippling of his father. Hidden behind a wall, he watched while a group of German soldiers beat his father with rifle barrels.[10] What he calls 'being terrorized' remains with him in his dreams and thoughts; it is there in his eyes, as he, with considerable difficulty, speaks about his past. He still bitterly remembers the collaborators and their role in the murder of his parents. Aaron's father, paralyzed by the beating, lay on his back for ten days until he and his wife were taken away by the Germans. Aaron found refuge with a sympathetic neighbor and eventually his brothers took him and his sister's family to the forests. It is not a time he wants to recall. 'Look outside, this Palm Beach, it is beautiful, heaven on earth. But I will try to tell you what I can.' Going back is not where he wants to be and the interview proceeds in fits and starts. But he insists we continue and tells me not to hold back on my questions. He will do as well as he can. What preoccupies him in our interview are the collaborators; he repeatedly comes back to them and their role in murdering Jews:

'I hated them, the father and son who worked for my parents; both took pleasure in seeing them trucked away by the Germans. My mother, someone later told me, asked the son for her galoshes; and he said, "You won't be needing galoshes where you're going." When my brothers several months later went into town and brought him back, I felt good when the fighters executed him.'

I asked Aaron if he would consider this a form of 'goodness' in the forests. With no hesitation he replied: 'Of course, when they killed him, it lifted the spirits of our whole group!'

A neighbor had informed on Aaron Bielski by pointing him out on the street to the Germans; he had been walking on the sidewalk, while his mother, because she wore the Jewish star, had to walk in the gutter. Arrested and interrogated as to the whereabouts of his brothers, this eleven-year-old was taken to the police station and forced to dig a trench. The police and Germans threatened to shoot him unless he told them where they could find Zush and Tuvia. The Germans ordered him to jump in the trench and lie down. 'They said to me this is how it will be when we shoot you. It was terror; I grew up very fast during those moments; but I told them I had no idea where my brothers were.' The terror, which he refers to throughout the interview, surfaces in his words, movements, pacing back and forth across the kitchen, reaching for his cigarettes. Even now, this 72-year-old survivor seems close to the terror, to its grip on consciousness and memory. It is like, as he put it, a 'cloud in the soul' that refuses to disperse. As the youngest Bielski, he had it 'pretty good in the woods... . I never had time to think about my parents.' Fighting and surviving defined his moral universe; 'I quickly became a man'; and while he had a rifle and gun, he never used them. 'But I wish my brothers would have let me kill Germans; I wasn't a pushover; when Germans or the anti-Semites who informed on us were killed, I wanted to be there; I wanted to see it; and there were times when I wished I had been the one pulling the trigger. But my brothers never put me in harm's way.'

In 1948, Aaron joined the Israeli army:

'They thought I was crazy, every night around 4:00 a.m. the Arabs screamed, "Kill the Jews, kill the Jews." ... I laughed as loud as I could. Compared to what I had been through, this was nothing. Don't get me wrong; the Israelis were great soldiers, as tough as they come. But no one in that unit had seen what I saw or survived with a partisan brigade.'

Relief from his terrorized consciousness came through the witnessing of violent retribution. He remembers an execution which helped his effort to forget, at least for a moment, the memory of his murdered parents or the memory of his father being beaten. A fighter in the unit who escaped detection in the ghetto by hiding in a cesspool 'up to his chin' for 48 hours, killed a collaborator:

> 'The man threw this collaborator down onto the floor; screamed this was for his parents, took an ax and chopped off his head. He and a couple of the others took the body and head and threw them on a bridge they had burned a day or two earlier. The note they placed on the body said something to the effect that this will be the fate of collaborators and anyone who turns in Jews.'

The stories increase Aaron's agitation; and after describing this scene, he searches for other images and memories, each one evoking the next, until they come spinning out of him in staccato words and hard glances. His childhood stolen from him, brutally buried in Novogrudek, Aaron still feels the hate. 'I can't get away from it; the death of my parents. You have no idea, the beating of my father, part of my life was formed when I saw him suffering and bleeding.' His innocence as a child had been shattered: 'My life was forced on me; I had no choice over this.' But his will as an adult recovered itself in the brigade. 'I had choices when I went into the forest.' What never disappeared was the rage: 'My anger never left me; it stayed inside, like a stone locked up in my soul. I couldn't cry about any of this until twenty years ago ... those few weeks in the ghetto, the police station ... turned me into ice. The forests protected the ice, but never dissolved it.' His brothers taught him how to control his fear. 'It was terrible, this control of fear; you have to put yourself in another place; you have to will to forget loss and pain... . We all controlled it, because if you didn't, you couldn't survive and you made mistakes. And mistakes meant death.'

When Aaron asked his brothers, Tuvia, Zush and Asael, why they had fled to the forest, all gave the same answer: 'It was simple common sense; what else were we to do?' But it was not simple common sense, certainly not to the tens of thousands of Jews in the ghettos who saw the forests as death. Aaron: 'We were never going to leave the woods; we would be dead before they put us in a concentration camp.' Yet, many Jews, including thousands from the villages around Novogrudek, saw it the other way: fleeing their families would bring death. Sonia Bielski understood their dilemma: 'Where would they have gone? They had no experience in the woods; if it's a choice between the certainty of staying with your family and trying to help them, and the uncertainty of a world you know nothing about ... where's the choice going to be?' The journey from ghetto to forest, filled with danger, required a resolve to leave family and friends and venture into a universe most Jews could not imagine. Aaron: 'Older people, city people, even people from our own town, made mistakes; they thought they could bargain with the Germans.' The *Judenrat*, he argued, rarely encouraged exodus from the ghettos: 'They believed the [Germans'] promises of resettlement. But even if there had been no collaboration between the Germans and the *Judenrat*, what did city people know from the woods? We lived there all our life, so the natural thing to do was to hide and live in them. We knew how to do it.'

But it would be wrong, he cautioned me, to think that the Bielski unit never had to deal with serious internal conflicts: 'It was not easy; there *was* jealousy and stupidity. But even though many people disliked others and fought amongst themselves, petty stuff, we all managed to cooperate; there were rules, for example, no rapes. We lived as a community.' Occasionally, city people found their way to the brigade:

'If city people found us, they adapted. I remember a barber came to our unit; he knew nothing about the forests. At first it was funny; he tried to climb up on a horse with his right foot on the left side. Of course, he ended up backwards. We laughed; his predicament and embarrassment lifted our spirits, if just for a moment. But after a couple of months the guy turned into a great fighter and went on missions with Zush's partisans.

We tried to lead a civil life; we had concerts in the woods, entertainers, like accordion players. And sometimes those concerts made us cry ... because they brought back the past.'

I asked him how he perceived his Jewish identity:

'I saw it through the anti-Semitism; the hatred; the anti-Semitism of the Soviet partisans, the collaborators, the Poles who wanted to kill us. That made my identity as a partisan, a Jew; it's why I took risks. I knew myself as a Jew who fought people who wanted to kill me. I was proud to be a Jew because the anti-Semites wanted to kill me. Look, here in the woods, I have my own rifle; I am surrounded by rifles. I know my brothers kill collaborators and Germans; I hear the stories; I speak with them after their missions.'

Tales of battles and victories restore the self; set right the past. 'You realize your pride; you're not afraid anymore; you look for the guy who beat up your father; you take revenge. That was our religion. It's different now, of course. I go to *shul* and I'm deeply religious; but then religion meant killing Germans and collaborators. It was a primitive life, but a proud life ... you see things and you understand what had to be done ...'

At that moment his voice trails off, and again, like so many of the partisan survivors I interviewed, his eyes, his consciousness, seemed to be back there and the tears start to flow:

'You asked me about religion, spirit ... those days I thought not about God, but about death, my death, my brothers' death, but even more than those fears, the deaths of those I hated. Some of the older people prayed; but I prayed that my brothers and his fighters would kill people. How could you not pray for killing when you saw what I did ... it was horrible.'

He stops abruptly; lights another cigarette, pours himself a small glass of whisky. 'I don't know how I can say this ...' He paces, but looks at me; I think, does he want me to ask him to stop, not to have to go on? But as I'm about to respond, he dismisses me with a gesture of the hand as if to communicate 'No, don't tell me not to speak.'

'You need to hear this ... a little ten-year-old girl is walking on the street; a normal day; I know her, a Jewish girl in the village.

Two Germans come up and start harassing her; they grab her, one takes her by the arm and the other by the leg; they laugh and pull, each one laughing harder than the other, each one pulling, straining. They pull until ... until they tear her apart ... she becomes pieces strewn on the ground. And the Germans walk off, still laughing ... . How, after seeing this, could I be religious? How could I not hate? I was a kid; these memories never leave me. My brothers and I, how could we speak about God? Would God let a sweet, innocent child be torn apart?'

* * *

As we stand in front of this lovely apartment building in Palm Beach, with its shiny nameplate and understated 1930s elegance, saying goodbye to one another on a warm soothing day, Aaron Bell takes my arm and looks at me, eyes deep blue, piercing, cynical, despairing: 'Thank God, you never had to go through these things; you grew up in heaven, with luck and luxury; you have no idea what it's like to be a Jew in a Gentile world.' Of course, he's right; what can anyone who has not been through this say? For the young Aaron Bielski, fighting kept him alive; but something else sustained his life, something more elusive, but still present in his words. Fighting, yes, kept him close to his brothers and partially blocked the images of his father being beaten and slowly dying. But his hatred, his witnessing and prayer for revenge suggest an effort at purgation, at ridding the self and its history of the omnipresence of death. Violence for the child evolved into a complex ritual of healing, a moral and psychological process that pushed back into the recesses of consciousness the direct memory of brutalization. But it seemed to me now that Aaron Bell was telling me that it failed; no matter how brutal he could be, how vicious in retribution his brothers and their fighters, the memory of violent loss never goes away, etched into the psyche like an ancient stone carving, an icon, guarding sacred terrain. Death never deserts memory. He wants me to tell the story; get it right, describe a terrorized child surviving the madness, but a survival riddled with wounds that lay open to time and consciousness. 'The terror; it is there, inside me; it is me.' It was that thought that stayed, circling in my head, a hammer shattering the peaceful landscape.

4
The Moral Goodness of Violence: Necessity in the Forests

Political organization and political action

Jewish partisans, tied together by the passions of revenge and hate, refused to allow themselves to be defined by despair; yet, they possessed an overwhelming advantage over ghetto inhabitants: partisan units never suffered the devastating hunger or the susceptibility to disease that afflicted the ghetto. Even though the mortality rate in some units was as high as 75 percent (most the result of wounds sustained in combat), the accounts of partisan fighters show individuals not only recovering their selfhood, but discovering forms of political organization that, while based on strict command structure, never existed in order to give over to the enemy a quota of bodies. Partisan units sought to engage the enemy and take the enemy's life; revenge, and in some instances rescue, not uncertainty and the hope that 'negotiation' characteristic of ghetto leadership would end the killing, drove these units.

Partisan brigades, both Soviet and Jewish, could hardly be described as democratic; the command organization derived from the Soviet army and communist methods of organization, and the brigade commander was an absolute ruler. Some underground units in the ghettos were organized as cells comprised of four or five individuals, each cell never knowing the identity of the others, with strict secrecy over membership defining inter-cell communication, and only commanders knowing all the details. Approximately 6 percent of all partisan fighters were Jews, but the proportion of

Jews engaged in partisan resistance to their own population was almost double that of other nationalities.

In the forests, some of the units were as small as 10–20 strong. By July 1944, one unit numbered 137 male combatants and 422 women, children and elderly and in the case of the Bielskis, as large as 1,200. All units possessed strong and inviolable lines of authority. Smaller family camps operated on similar principles. Women, children and the elderly functioned in support capacities; combat assignments were for those with weapons. There were some eight to ten family camps in Byelorussia. These family camps played a major role in saving the lives of thousands of Jews who otherwise would have perished in the forests; approximately 5,200 Jews found their way to them. The casualty rates were low in these camps; the Bielski Brigade suffered only 6 percent casualties, as compared to 75 percent in some fighting units.

One partisan recalls that, after reaching the safety of the forest and joining a local partisan brigade:

> 'A remarkable transformation took place. We felt stronger, almost unconquerable. In possession of an instrument for defense and attack, we had absolute control. Only yesterday we were creeping through the fields and villages, dreading every rustle and movement, fearful of the light of day. As an organized fighting group, with these weapons, we were able to leave the forest on horseback or on foot, with our guns suspended from our shoulders.'[1]

On balance though, even in the face of knowledge about the transports and the gas chambers, partisan bands had little influence on ghetto inhabitants. A letter smuggled into the Stolptiz ghetto by a partisan courier reads:

> 'Time is your enemy. Organize yourselves while it is still feasible. Your young lives are precious. Let them not be destroyed. At all cost collect weapons. Time is short. Prepare to join in the partisan life. We can no longer listen to what our elders say. We must fight. I speak from experience. Nothing will be gained by our deaths in the ghetto.
>
> Your friend, Siomka Farfel'[2]

If Frantz Fanon had known of these brigades, if he had met Zush Bielski, he may very well have used the data in *The Wretched of the Earth*[3] to bear out his thesis: violence restores the psychological health of the oppressed and victimized. For Fanon, the colonized recovered their humanity only after they had crossed the psychological divide separating inaction, passivity and acceptance from action, transcendence and negation. Violence not only negated the oppressor's effort to dehumanize the victim, it also did away with the negative self-image imposed on the victim by the oppressor or colonizer. In the case of the Jews, being outside the ghetto and the deteriorating environment of compromise, death and collaboration, the self could begin, once again, to live as a human being, but the action of living required the externalization of rage, turning against the aggressor what the aggressor had used against the self: violence.

Survivor accounts of killing Germans, in working together without fear of degradation, suggest that traditional values of compromise, negation and reason – used by the *Judenrate* with the Germans – could not work. Partisan leaders like Zush Bielski refused to compromise; they acted decisively and made a virtue out of killing (it was the supreme value in the camps) – a perspective beyond the vision of most *Judenrat* authority. In the forests, killing restored the self's humanity and brought back from the dead many who psychologically had been shattered by ghettoization. Faye Schulman recalls: 'My life consisted of swamps, water, ice and freezing cold weather; my lot a rifle and a bed on the hard ground winter and summer. I was not allowed to be sick – there was no medicine. The water I drank was full of bugs, but nothing happened to me. I was stronger than steel.'[4] The forests brought her life.

A woman partisan remembers the pleasure of violence:

'Everyone started beating them – with rifle butts, fists, boots. We beat them to mush. I remember that they were lying on the ground just barely breathing. And I ... I don't think I could ever do it again... . I came up to one of the German officers who had his legs spread. I started to kick him again and again in the groin. I was kicking and screaming, "For my mama! For my *tate* [daddy]! For my sisters!" I went on screaming out every name I could remember – all my relatives and friends who had been murdered. It was such a release! It was as if I had finally done what my mother had asked me to do. I still remembered my mother's last words as she was waiting to be taken to the grave, "Tell Rochelle to take *nekome* [revenge] – revenge. Revenge!"'[5]

The damage done by Jewish partisan units including the Bielski Unit suggests their success in implementing the philosophy and action of violence. The Vilna detachments alone derailed 242 trains, 113 locomotives and 1,065 railroad cars; destroyed 12 storehouses, 35 bridges, 257 vehicles, 1,409 miles of railroad tracks, 4.2 miles of communication wires and 11 tanks; and engaged over 4,800 enemy soldiers.[6] What is so striking about these figures is how strongly they contrast to the despairing diaries typical of ghetto life, and how critical the issue of violence was to Jewish survival and the recovery of Jewish self-respect. The following is an eyewitness account of partisan actions against the Germans; it is typical of what the fighting units accomplished:

> 'I belonged to the sapper commandos. I crawled under bridges and planted mines. We were followed by an inspection unit checking to see if we'd carried out our assignments. German trains crashed down off the banks or were smashed on the tracks. We captured tons of supplies headed for the front. I manned a machine gun and stopped the Germans from harvesting the wheat.... . I belonged to the assault group. There were seventy-five of us. We blew up the three German artillery pieces... . Several Jewish and Russian soldiers and I liberated a transport of 400 Jewish girls, who were barefoot and almost naked... . The girls were in a pitiful state. Some muttered as if they were insane.'[7]

When the Germans met resistance they backed off, whether that resistance came from people like Raul Wallenberg in Budapest or the five-man partisan brigade deep in a Russian forest. Disconnected, fragmented selves, falling into a deadly catatonia (common in the ghettos) were much less likely to appear in resistance bands – and much of the explanation for this can be found in the *group* project whose primary objective lay in warding off threats to identity through the organization of violence and the maintenance of communities committed to violent action.

Violence, then, in the administration of these political bonds took on a powerful valence of goodness. It possessed an ethical component: killing was a requirement for membership in all partisan units. No member of a partisan unit, whether the family units or the fighter-only brigades, whether Jew or Soviet, regarded violence as morally inappropriate action. Informer Jews were ruthlessly executed. If killing an informant could save a Jewish child, then violence became a holy duty, an act of moral goodness.

In the family bands, children for the most part collected berries or mushrooms and performed odd jobs around the camp, but some did fight. Children grew up quickly in the forests. An eleven-year-old partisan fighter reports the following: 'Once, Kolodko, the village chieftain of Zaczepice, informed on us. We found out about it. We went to him, supposedly on a "courtesy call," and shot him dead. We avenged ourselves like this against five peasants who informed on us.'[8]

The more the partisan units harnessed violence to the rescue aims of the group, the more the group succeeded in its protective functions. The more vicious the band towards informers and collaborators, the easier it was to enlist support from the peasants for supplies. The more violence was used against German patrols, the less likely it was that the patrols would undertake search-and-destroy missions. The more ruthless the fighters were towards those who killed and tortured Jews, the less likely were others to engage in similar behavior. Partisan violence never produced indiscriminate consequences; it focused on specific ends. It had a limited aim, very much *unlike* German violence, which destroyed anyone in its way, including the innocent.

Partisan political organization had as its fundamental guiding project the pursuit of violence; but a discriminate violence, to be exercised against those who had transgressed the line between barbarity and civility. In the case of the Bielski Brigade, violence was pursued concurrently with the rescue of anyone in need of shelter from the brutalization of the German assault. In the ghettos, it had been a totally different story. The *Judenrate*, for the most part, encouraged withdrawal from violent engagement; they preached a disdain for action and a belief that 'reason' and self-interest would eventually triumph; that if one sat down with German officials, certain arrangements could be made. Strong arguments in the ghetto opposing this view rarely were accorded a public forum. Partisan and underground leaders understood, if not explicitly then certainly intuitively, that in the world the Germans inhabited, it was not reason that defined moral order and law, but a psychology of domination intent on expunging all impurity – and that meant the Jews – from the universe.

It is not that violence was redemptive in any religious sense; rather, violence reclaimed the self from a shattered psychological universe. To crave violence, to pursue violent reprisal, meant the self restored to itself faith in human possibility, in the value of community. Killing Germans and collaborators simultaneously affirmed community and what community signified politically and socially: a home (even if it was no more than a hut); crude workshops, meager but adequate food supplies; herbal medicine (root and animal extracts); some faith in the future and trust in the present. The social activity characteristic of partisan life, supported by a politics of violence, depended on human beings demonstrating a faith in life itself – a faith that had been seriously damaged in the ghetto by German brutality. Not to live in fear of imminent starvation; not to live in fear of being discovered in a cramped, filthy, hiding place; not to live as abject, as downtrodden, but to live as a sentient, active, emotion-filled self, producing moral recognitions in which killing Germans meant not transgression but an affirmation of history, tradition, culture and, most importantly, the future of the children – this is what the forests brought to Jews. In the Bielski camp children were taught about the world, not with books but with teachers telling stories, singing songs, going on excursions into the forests, and encouraging children to be children once again instead of beggars, thieves and orphans reduced to despair, madness and hopelessness.

The family camps accepted children but discouraged pregnancies; harsh living conditions, severe weather, the uncertainty of German and collaborator raiding parties, made an environment hostile for infants. Abortions were common in all the family camps, although a few infants were born there. As one Bielski survivor said to me: 'Our *yishuv?* We built the book of survival through the barrel of a gun ... but we also performed many abortions. We gave life and took life.'

It was not that inspiring, charismatic leadership had no presence in the ghettos. It was there, but hidden, in the undergrounds, away from the community, disguised in proclamations and leaflets but rarely encountered in face-to-face relationships. The covertness of this kind of leadership, its inability to collectively motivate the ghetto, can be traced directly to the *Judenrate* and their principles of survival. If the underground had been permitted to address the ghetto collectively, if their arguments had been debated publicly in ghetto town meetings, if the *Judenrat* 'rationality' of saving the remnant had been subjected to critical, public inspection, then the undergrounds might have had audiences of more than a few hundred; they might have been able to work with the traditional leaders on techniques to resist the Germans; they might have placed the philosophy of saving the remnant before the community, even though most Jews were predisposed to inaction. But with very few exceptions, *Judenrate* and undergrounds and partisans worked at cross-purposes, a point consistently argued in underground pamphlets and publications.

Underground leaders, at the time, were acutely aware of the failure of traditional leadership. The unwillingness, then, of the *Judenrate* to admit the undergrounds to their governing councils was indeed a failure of leadership and a failure of authority. This is not a question of the benefit of hindsight; it was an issue apparent to what the undergrounds consistently wanted, what they publicized in their leaflets, in their incessant pleas to the *Judenrate* to listen to their arguments, to allow them to publicize German atrocities. That the vast majority of the *Judenrate* chose to collaborate rather than to listen to their own people, to the most courageous leaders in their communities, undoubtedly had a considerable impact on why millions met their deaths worn out from the very struggle with life.

Violence and the recovery of self

For Frantz Fanon the oppressed calls his world into question through the use of 'absolute violence.' It is the human and psychological condition for the recovery of identity. 'For if, in fact, my life is worth as much as the settlers, his glance no longer shrivels me nor freezes me, and his voice no longer turns me into stone. I am no longer on tenterhooks in his presence.'[9] Much the same can be said for resistance and partisan groups of Jews during the Holocaust. In looking at Fanon's theory of violence, his concept of the colonized becomes, in the context of the Holocaust, the 'being-ness' of the Jew, the oppressed, the exploited, the 'wretched of the earth,' which the Jew was for the German.

In an extraordinary passage from Aimé Césaire's *Les Armes Miraculeuses*, Fanon locates a moral story in the oppressed's redemption through violence: 'There is not anywhere in the world a poor creature who's been lynched or tortured in whom I am not murdered and humiliated.' In the next passage, the rebel enters the master's house: 'The master was there, very calm ... and our people stopped dead... . it was the master... . I went in. "It's you," he said, very calm.' But the rebel saw fear in the master's eyes and he 'struck ... and the blood spurted; that is the only baptism that I remember today.'[10] Analogous accounts appear in underground and partisan memoirs; the Jew, striking, killing a collaborator, and the sense of exhilaration and rebirth such action provoked.

Violence is reciprocal; it feeds off itself; when the oppressed find their voice, their violence matches that of the oppressor. It is messy, but what distinguishes the oppressed in their violent phase is their sense of purpose, what Fanon calls the 'point of no-return,' when the first act of killing marks off the rebel as actor from the oppressed as passive recipient of violence. At that point, 'violence, because it constitutes their only work, invests their characters with positive and creative qualities'; violence adheres with identity, recovers selfhood, while 'the practice of violence binds them together as a whole, since each individual forms a violent link in the great chain.'[11] The Jewish underground fighter initiated into the group, the relationship between the fighter and his weapon, the Bielski band's sense of great victory at the successful expedition of its fighting units, the accounts of partisan fighters filled with pride after having engaged in a violent action against a German or collaborator; all this evidence suggests a positive correlation between the practice of violence and the recovery of a vitalized community. This certainly was Fanon's belief in relation to the Algerian peasants' subservience and victimization at the hands of the French. 'Violence is a cleansing force. It frees the native from his inferiority complex and from his despair and inaction; it makes him fearless and restores his self-respect.'[12] And it was a recurrent theme in my interviews with partisan survivors.

The eleven-year-old Jewish boy who kills his first German feels a sense of victory and vindication at the very act of destruction. No guilt or remorse appears in partisan accounts, although survivors like Sonia O. continue to struggle with the moral implications of group murder and execution. In a videotape of surviving members of the Bielski Brigade, one man says quite matter-of-factly that they had to kill, that it was a part of their life, but in his eyes and his wife's you witness a sense of vindication and pride at having encountered the German oppressor on the ground of violence. For Fanon, 'life [for the colonized] can only spring up again out of the rotting corpse of the settler,'[13] for the Jew, out of the rotting corpse of the German. The oppressor divides the world into the good and the evil; 'the Manicheanism of the settler produces a Manicheanism of the native.' But this is all to the good, because it incites resolve with singularity of purpose; it dissolves the sense of limiting action and replaces a cautious model of action with a radical one. In Fanon's words, 'To the theory of the "absolute evil of the native," the theory of the "absolute evil of the settler" replies.'[14] To the absolute evil of the Jew, the Jew replies with the absolute evil of the German, Nazi collaborator, turncoat; action as violence realizes in practice this Manicheanism carved into the spirit. Rather than degrading the self, this purposeful violence fulfils the self.

The Germans, to use Fanon's words, had taken away 'the warming, light-giving centre where man and citizen develop and enrich their experience in wider and still wider fields'; for Jewish resistance fighters, violence brought that experience back, enabling the community to exist and survive. When the masses, Fanon argues, give 'free rein to their bloodthirsty instincts,' action, rather than ruining character, restores it and gives it purpose and a sense of place in a tormented history and potentially redemptive future.

Oppression kills the self, maims human motive and distorts the natural aggressiveness necessary to engage in violent action. However, the stunted self, the self shorn of its human properties, the dissociated, alienated self, literally reverses itself through violence; and in the process of coming together as a revolutionary community, the oppressed discharge 'the hampered aggressivity' and destroy oppression 'as in a volcanic eruption.'[15] The result is catharsis, rebirth, a psychic moving outwards in a burst of energy and rage that brings back to the group its human and political identity. In violent action, 'the evil humours are undammed and flow away with a din as of molten lava.'[16] But in captivity, in oppression, 'the native's back is to the wall, the knife is at his throat (or more precisely, the electrode at his genitals).'[17] But when these fears dissolve in action, 'a people becomes unhinged, reorganizes itself, and in blood and tears gives birth to very real and immediate action.'[18] In conditions of oppression, 'the psyche shrinks back, obliterates itself and finds outlet' in behaviors which work in the interest of the oppressor (group infighting, the use of drugs, theft, black markets, intra-group exploitation).

The colonist tries to break down community and enforce a regime where individuals find themselves locked into their own fear, their 'own subjectivity.' This produces isolation, makes it difficult to transcend or break through domination. Action obliterates this paralysis, 'where even in its own universe, amongst its own people, each self is enemy to everyone else.' Instead, what revolutionary violence accomplishes is a reinforcing of the consciousness of community, 'brother, sister, friend – these are words outlawed by the colonialist bourgeoisie.'[19] Through violence, the political meaning of these words is rediscovered, and all in the community come to see themselves as common participants – not isolated units trying to survive at all costs, a state of mind typical in Holocaust diaries describing the horrifying conditions of ghetto life: 'So when I as a settler [Jew] say, "My life is worth as much as the settler's [German's]," his glance no longer shrivels me up nor freezes me, and his voice no longer turns me into stone'[20] – for example, the difference between the deliberations of the Vilna *Judenrat* and the Bielski partisans, the former reinforcing the power of the German to turn the Jew into stone, the latter, confronting the German presence with a massive negative of the effort at petrification. Fanon's natives 'live in the atmosphere of doomsday';[21] so do the resistance fighters, a sense of imminence defined by the implacable hostility of one group against the other, except that, for the colonized, 'doomsday' means not doom for themselves but for the colonialist. The native's work is to imagine all possible means of destroying the settler. So too for the resistance fighter; no compromise is possible. And in an observation that could have come from Tuvia Bielski, Fanon observes:

> 'The practice of violence binds [the community] together as a whole since each individual forms a violent link in the great chain, a part of the great organism of violence which has surged upwards in reaction to the settler's violence in the beginning.'[22]

Fanon refuses to leave the argument at the level of violence alone; he realizes the critical importance of the formation of community accompanying the practice of violence. This is where, for example, Tuvia Bielski's theory of the absolute necessity of a Jewish community as the backdrop for the fighters bears a fascinating resemblance to Fanon's arguments. Violence by itself may be cathartic, but it lacks political force and focus; community gives violence its structuring properties, while containing its effects. The practices of communal exchange and rebuilding that follow on violence guarantee the community's future as a political entity. Violence creates the possibility for a regeneration of community; it is not a substitute for it. Bielski constantly made this argument not only with his own people, but with the Russians, their commanders and with other partisan units. In Fanon's words, 'The leader realizes, day in and day out, that hatred alone cannot draw up a programme: You will only risk the defeat of your own ends if you depend on the enemy (who of course will always manage to commit as many crimes as possible) to widen the gap and to throw the whole people on the side of the rebellion.'[23] Hatred requires a process of *structuring in community*; of harnessing action/violence components and transforming those components into a community that lives for today and tomorrow. This is what imparts to violence its political meaning and practice. What was remarkable about the partisans was how political they could be when necessary, when the metamorphosis of violence into politics – for example, Bielski dealing with the Russian commanders – meant saving the integrity of the community itself. Uncontained, cathartic violence for Fanon could not sustain revolution; it had to be transmuted into structure and communal practice; and that transformation is what gave the rebels strength in their political confrontation with the French. Similarly with the partisans: what enabled the resistors to survive in the loose collective of Soviet-sponsored bands was not only the strength of these fighting units, but the demonstrative power of whom they represented as community, a fighting entity surviving in the midst of German annihilatory violence.

Perhaps the violence of the oppressed blurs the line between reason and madness, or perhaps the madness of what happened to the Jews, as incomprehensible as it is, allowed them to forge out of the forest and undergrounds a psychological space of action allowing for the interpenetration of reason and madness, providing moments of lucidity in a human condition where it was all too easy to be driven insane by genocide. It was certainly the case that many Jews were driven mad by German barbarism. And the ability to maintain one's sanity, to avoid falling into the stupor of catatonic withdrawal or the passivity of psychological dissociation, required containment by action offering to the alienated hope and possibility through violence.

There is an account in the diaries of the Lodz ghetto of a woman released from a mental hospital (or a hospital ward housing the insane) who went to a guard post and asked the sentry to shoot her. He obliged, but not before he insisted she dance for him. Madness, lunacy, appears in many different forms in Holocaust ghetto life: in the blank stares of lice-ridden children, sitting in gutters filled with human refuse; in the empty faces of those starving to death, those who had lost entire families to the Germans; in the wailing of orphans desperately looking for their parents; in the catatonic withdrawal of adults who lost the will to live. Possibly the violence of the resistance movements served as moments of lucidity or reawakening to one's humanity and willfulness, in the midst of a universe thoroughly turned inside-out, dedicated to destroying the will. Violence kept resistance fighters from falling over the edge into incoherence or madness; the undergrounds, the partisan groups in the forest, were composed of human beings who had been driven to distraction by loss, suffering and pain, but who nonetheless managed to transcend the indignities imposed on their bodies, families and minds by the Germans. Violence did bring lucidity, a kind of reasoning born of circumstances certainly not normal in our understanding of 'normal.' We do not regard it as healthy to engage in acts of murderous violence, but in the madness created by the Germans, maybe the resistance unit constructed lucid intervals filled with clarity over one's meaning as a human being and the obligation one had to support the community, a lucidity that paradoxically allowed individuals to regain their humanity, will and sense of a common purpose. It would be wrong to speak of this kind of violence as normal, but in the context of that time, it did work, it did bring lucidity and hope, and for many, it enabled them to survive: 'the partisans were hunting down the Germans. Our spirits were high. We were enthusiastic and vengeful, glad to batter the enemy.

Every evening, when we returned to the base, we spoke excitedly and joyfully of the day's events.'[24]

The clarity of the need for violence appears in the narrative of Ben, who escaped the Warsaw ghetto when he was seventeen and joined a loosely organized partisan group outside Lublin. The head of this unit had no use for Jews, and made this clear to Ben:

'Several Jews managed to escape a transport to Maidanek; they had some money on them; and my commander killed all sixteen of them. Just like that without warning before we could react; shot them for the money, the gold. We couldn't believe it, right before our eyes, my friend who was Jewish and myself, we just stood there for a moment, looking at each other. We pulled out our guns and killed him; must have pumped ten shells into his body. I knew what my mission was: to kill all those who killed the Jews, even partisans. I hated that guy.'

Ben, in a story unusual for its absence of anti-Semitic harassment from his comrades, fled eastward and joined a Soviet partisan unit.

'They were big, and you know what? They didn't mind I was Jewish; no one in that unit hounded me. All they cared about was that I killed Germans. I know you find this hard to believe, but the Russians never called me Jew or anything like that. I was treated like any other fighter. And I did fight; I killed many Germans and blew up a lot of trains.'

Ben's unit operated with the support and assistance of the Soviet government; weapons were airdropped, and they encountered no problem in intimidating local peasants into supplying food. 'If the peasants didn't fork over food, we beat them; sometimes we had to kill them. But it didn't matter; we had to have food; the peasants were scared of us. If they betrayed us, we would kill the collaborators. They learned quickly.'

Life in the forests possessed no routine.

'Every day there was something different; you couldn't dwell on yesterday; you never knew when a local collaborator would take a shot at you or a German unit on a sweep would ambush the unit.

We had to survive; six guys had one large pail they ate from; we threw whatever we had into the pail; if there wasn't enough we would try to find a peasant; if the peasant refused us food, we threatened to kill him; that was usually enough. During the Christian holidays, the peasants baked: cakes, bread, cookies; we took whatever they baked! And they never knew when we were coming; we ate good during those days. What really bothered us was the lice; we couldn't bathe or wash; we wore the same clothes for months; all of us were full of lice; little bugs, everywhere, our hair, clothes, body.'

Ben's unit had the specific assignment of destroying rail lines that carried armaments from Germany to the Eastern Front. He remembered how successful they had been, and the pleasure it gave him to blow up a train and

'kill everyone on it and steal what they had ... that was a terrific feeling, to see the bastards blown up, and to shoot the stinking Germans. I had no regrets about killing them; I regret not having killed more Germans. We especially liked to rob the trains at Christmas time with all the gifts going to the front ... anyone who tried to hurt us. I hated them. After the war, Poles would hurt us; I never helped them; I gave them poison [a colloquialism for wanting to hurt those who hurt you]. Some Russians helped me but I didn't look Jewish to them and they never asked. The Poles, they would ask.'

In postwar Poland, the Soviets appointed Ben chief of police in a small town in eastern Poland; he sent forty Poles to jail for crimes against Jews. 'The Poles didn't have an order to kill the Jews; they did it on their own. So in a way the Poles were more vicious than the Germans; they chose to kill.' His bitterness towards the Poles appeared consistently throughout his narrative:

'Sometimes during the war I had to hide my Jewish identity; what else could you do? Even after the war, when the Poles found out I was Jewish, they screamed that a Jew should not be chief of police and send Poles to jail. Even in that small town I lost two friends who were killed on the street by Poles. After that happened, I said, "To hell with it; I don't want to stay here." And I left and came to America with five dollars in my pocket.'

In the ghetto 100 percent of deaths came from disease, starvation and killing by the Germans and their sympathizers; in the forests approximately 72 percent of all partisan deaths came from combat; 3 percent were caused by disease and the rest at the hands of hostile, anti-Semitic partisan bands. Men composed 80 percent of partisan units, women 20 percent, although some family brigades had a greater proportion of women, children and elderly in their populations. The vast majority of participants in the ghetto underground were under 30 years of age, with the majority of non-command fighters being 20 years and younger. While it is difficult to assess with any precision, the ages of those in the forests were generally higher, command fighters and leaders being 40 and under; non-command fighters, anyone who had the strength to endure combat conditions and who possessed a weapon, were for the most part under 30.

Dov Levin[25] counts 5 percent of the Jewish population in the eastern areas of Lithuania and Byelorussia involved in political and active resistance; and in Lithuania as a whole, Levin estimates that 2,000 Jews served in the underground or with partisan units. Yehuda Bauer argues that in eastern Poland 'during the time Jews were organizing to fight, that was in 1942–43, they accounted for one half of all partisans in the Polish forests, or about 1.5 percent of the Jewish population in the region.'[26] As a proportion of the *entire* Jewish population, this figure is higher than equivalent numbers of Polish and Lithuanian resistors. Because captured Jewish partisans were tortured and killed most tried to commit suicide or kill their enemies. One young woman blew up herself and a German officer with a hand grenade. On occasion wounded comrades were put to death before capture by the Germans.

A partisan attack killing two German officers provoked the following proclamation by the German command in Vilna:

'Men and Women of the Vilna District!
400 saboteurs and terrorists [were] shot in reprisal... . We call upon the population of the Vilna District to rise against the Bolsheviks and terrorists. Report the presence of strangers in the

area at once. Anyone aiding the Bolsheviks and terrorist gangs, or failing to report, will be severely punished and liable to the death penalty... . All those imparting correct information will be highly rewarded with money or a gift of food.

                       Gebietskommissar of the Vilna Region WOLF'[27]

For ghettoized Jews, 'escape' had limited meaning: one could attempt to obtain a labor pass, forge a fictitious marriage or marry someone already in possession of a labor pass; one could hide in a bunker or a hole or what was called a *maline* in the ghetto; one could hide with a Christian family outside the ghetto; or attempt to flee to a neutral country (almost impossible after 1940), or more likely to the Soviet Union or for a brief period in Byelorussia, where Jews lived in relative safety until the German invasion. If a Jew had the right papers and physical characteristics, he might pass for an Aryan.

Trapped, without resources, starving, the majority of the Jewish population dealt with the German threat by hiding and waiting. Families were desperate to stay together. Those Jews young, strong and fit enough to escape to the forests found themselves in an environment where choice reappeared, where control over one's life took on political significance through the exercise of violence. Jews inside the ghetto who were fortunate to possess the infamous yellow or white passes exempting them from selections were required to carry the certificates with them at all times; to accept the work offered by Germans without question; never to change their place of employment without permission; to obey the orders and instructions of the ghetto's labor department; and to notify the labor department of a change of address or change in place of work.

Every aspect of life was governed by the ghetto and German authorities. No one could act without some form of permission, in addition to the obvious fact of being encircled by barbed wire and armed guards.

A German poster read:

'The Jew is the enemy of Germany and responsible for the war. He is a forced laborer and is forbidden to be in contact with his employers except on matters referring to work. Anyone maintaining contact with Jews shall be treated as if he were a Jew.'[28]

Escape to the forests

On September 1, 1943, German units entered the Vilna ghetto intent on what was believed to be a 'labor' expedition. Joseph Gens, the head of the *Judenrat*, had worked out an agreement: if German forces withdrew immediately, if they postponed their search for 'laborers,' he would supply the needed 'labor' to fulfill the quota. For the next two days ghetto and auxiliary police rounded up workers for 'transport' to Estonia. It was in effect a death sentence. The underground bitterly fought this collaborationist policy and lost. The underground organization FPO issued a manifesto:

> 'German and Lithuanian hangmen have arrived at the gates of the ghetto. They have come to murder us! ... Do not cower in the hideouts and *malines*. Your end will be to die like rats in the grips of the murderers. Jewish masses! Go out into the street! Whoever has no weapons, take up a hatchet; and whoever has no hatchet, take steel and cudgel and stick! For our fathers. For our murdered children! To revenge Ponar, hit the murderers!'[29]

It had been almost six months since the Warsaw uprising; a few hundred Jewish fighters held off the German Army for three weeks. The Vilna ghetto knew what the underground had accomplished in Warsaw; yet very few ghetto inhabitants, a few youngsters, joined in fighting off the German attack.

Shortly after this *Aktion*, the Vilna underground began to disperse and leave for the forests. Locating a partisan unit in the forest involved skill as well as luck: skill at surviving the early days of forest life; luck in evading anti-Semitic peasants and partisan units, and finding one's way to a partisan group willing to accept Jews. With the exception of some Jewish partisan bands, non-Jewish partisan units, usually headed by Russians and escaped Russian prisoners of war, if they accepted Jews at all, required that the Jew have a weapon. No Jew would be accepted into a non-Jewish partisan band without a pistol or rifle. If a Jew were fortunate and lucky enough to find his way to a partisan unit, survival required acclimation to an extraordinarily difficult life: harassment by German units and anti-Semitic police and para-military units; the need to find and extort supplies from unwilling peasants; harsh winter conditions;

shortages of medicines and bandages, and the anti-Semitic diatribes and rages of non-Jewish partisan fighters.

Partisans had to be inventive; for example, a group of young men having a few weeks earlier escaped from a ghetto, built a bunker in the forest and faced the task of finding supplies for survival. For days they begged for food but the peasants refused them.

> 'Sitting by the bonfire one night, a new idea occurred to Musio and me. I cut off the tops of my boots and sewed them into a holster. It took me a whole day to do it. Shaping a piece of wood to look like a Soviet Nagan revolver, which had a wooden hand grip, I stuffed it into the holster. It looked like an authentic revolver when I attached it to my belt.... . We would no longer beg for food. We would demand it!'[30]

His comrades made rifles out of sticks; the peasants, frightened by what they mistook for weapons, gave the group what they needed. Yet the deception was not easy: 'The negotiations were full of tension. Confident on the surface, I quivered inside. I worked out a system of acquiring as much as I could without overstretching the limits and causing rage.'[31] But eventually the peasants figured out what they were doing.

Partisan groups constantly changed composition, particularly those commanded by Russians; separation generally followed the lines of whether the members had a weapon. Those with rifles separated themselves from the unarmed, the elderly and women. The composition of groups was fluid; as soon as one had been organized, it was dismantled. Jewish refugees often banded together and formed units of their own and then joined other units. Eventually, in 1944, the Soviets insisted that all Jewish partisan units merge with Soviet groups and accept Russian leadership. Hundreds of Jews, fleeing burning ghettos or escaping certain death at the hands of Germans, roamed the forests, trying to survive in bunkers, begging. Ben: 'We found Jews hiding in the forests; our units took them in, gave them food. By the end of 1944, we had over two thousand Jews in our unit that we were protecting.'

Anti-Semitism was a force to be reckoned with: 'So here we were, fighting against a common enemy – the Germans, whose aim it was

to totally annihilate the Jewish people and to take over the Soviet Union – side-by-side with fellow fighters whose own hatred of Jews was notorious.'[32] In addition, partisans in the Polish national army (AK) would often attack Jews and Russian partisan groups with more fervor than their fight against the Germans. As resistance survivors told me over and over, 'The Poles hate the Jews as much as the Germans do.'

If it were not the Germans, the local police, the anti-Semitic partisans or treacherous peasants who killed Jews, it was disease or wounds. 'I had an abscess in my throat and a raging fever. Lying on my bunk, I was unable to move or eat… . I couldn't swallow or talk, and even drinking milk was painful. I had difficulty breathing and felt the swelling in my throat was choking me.'[33]

Jews in Russian units felt isolated. Yitzhak Arad writes of his experience:

> 'How could a Russian kolhoznik possibly understand the Jewish fate, the loneliness, misery, and ruin on all sides, the importance of saving every Jewish child in the face of the mass annihilation of our people? Both of us were partisans in the same unit, fighting a common enemy, but a deep abyss separated my war aims from his.'[34]

Even though he might die, he realized his death would be on his own terms and not the Germans': 'From this moment on my comrades and I were not humiliated Jews under Nazi rule, sentenced to annihilation, but free fighters who had joined the millions fighting the Nazi beasts on all fronts. I touched the revolver hanging at my belt and the grenade in my pocket. I felt great confidence. The ghetto was behind us, the forest and the unknown before us.'[35] In our interview, Arad constantly stressed how critical psychologically was the feeling that he could kill Germans.

Nahum Kohn, a resistance fighter in Byelorussia, describes what violence meant to the partisan:

> 'I will tell you about a world that went crazy, a world where humans became beasts, turned worthless, and the forest became "home." And I will also tell you about people who refused to surrender to bestiality, people who resisted the descent into

darkness. Most of them are gone, but I see them still, in their tattered rags, city boys darting from tree to tree in the forest, repaying death with death. I, too, was a city boy, and although I survived, I have never really left those forests.'[36]

What brought hope to these desperate fighters was the impulse to kill: 'Only one thing mattered: force, might';[37] to kill Germans, to take revenge on collaborators.

'We pledged that we would not sit on our hands in the forest; we would not steal food and hide; we would *take revenge* specifically against murderers and butchers. We would go after those people who had our brothers' blood on their hands ... [I] couldn't control my boys; one jumped on him [collaborator] with a stick, and another beat him with a piece of iron.'[38]

Possession of weapons meant everything; and the possibility of being able to exercise violence was transformed into a central part of one's humanity and survival. 'A revolver was gold, diamonds, *everything* – without it you were regarded as a cockroach. With it you became a respected person.'[39]

After Soviet control had been consolidated, Kohn returned to his birthplace, his hometown, where he learnt that all his relatives and family had been killed. 'I was like a lost sheep. The flock had gone in one direction and I had gone off in another direction, and I was lost. I used to cry like a lost sheep that bleats in its panic and solitude, all alone.'[40] If the self's civility had been restored in the forest, it faced terrifying challenges in the aftermath of the Holocaust. Discovering death, destruction, houses burned to the ground, parents, sisters, children, husbands and wives missing with no record of their death, survivors faced yet another set of challenges in addition to an anti-Semitism that plagued them even after the Germans had been defeated. In the words of Ben: 'I returned home; but my family all had been killed. I had nothing. I saw a bunch of Poles coming towards me, muttering, "Jew, Jew, Jew." I knew at that moment I couldn't stay. I had two feet to run away from that town. I will never go back.' Many resistance survivors found ways to emigrate illegally to Palestine.

Politics in the forest: community transforming self

It would be wrong to assume that partisan life should or could have attracted a greater percentage of Jewish fighters and participants. It took almost superhuman effort to reach the forests and the Jewish ghetto population was incapable of taking those steps. Family ties, weakness, starvation, German patrols, the difficulty in obtaining weapons, and, above all, the Nazi policy of mass reprisal made escape a treacherous undertaking. What the forest partisans demonstrate is the effectiveness of alternative leadership, the extraordinary importance of the concept and action of rescue.[41] It also reveals the vitality of communal values in facilitating the work of escape and rescue, and the relative weakness of traditional, *shtetl* values in sustaining resistance and rescue. In the Bielski Brigade, for example, while not a dictatorship (although Bielski's word contained the power of law; and his rule was undisputable), his leadership depended on the absolute recognition of his authority as the supreme representative of the community's will. The fighters and the support group, without hesitation, actively gave Bielski the power of sovereign authority. Conflict over leadership (in one case Bielski executed a person posing a serious threat to his position) endangered the welfare of the group, the community; to conspire against the leader or demand his removal fractured unity and therefore threatened the group's effectiveness as a place of rescue and support for refugee Jews. The community's civil religion, if one may use Rousseau's term here, defined itself in terms of both rescue and vengeance; therefore, challenges to leadership posed a serious threat to the group's venue and its practices. Certain kinds of dissent, particularly against the leadership and political decisions, had to be silenced to preserve the group's integrity.

Vengeance for most partisan fighters took on the status of a faith; it forged the group's ends and purposes, and the purpose of life itself. This, however, was not the case with the Bielski partisans, at least not Tuvia Bielski's primary purpose. Rescue, as political organization and partisan cooperation, molded values, and forest life proceeded around the reality, not fantasy, of rescue. Vengeance always was there; but it did not constitute the day-to-day policy. This is not to say the fighters never went on expeditionary or reprisal raids. Bielski's group killed when it had to, whether the victims were

Germans, collaborators or peasants who had tortured and killed Jews. The Bielski partisans, as skilled guerrilla fighters as any partisan band, pursued vengeance enthusiastically. However, from the very beginning, Bielski formulated his detachment's 'general will' as that of saving Jews; and rescue took precedence over all other factors – including vengeance.

Bielski's Rousseauian community[42] demonstrates how fragile traditional values in dealing with the 'outsider' are in an environment where the political and cultural rules of the game have been shattered. What today we take for granted – individuality, tolerance, skepticism, suspension of belief, respect for the rules of the game, and so on – in the ghettos and forests became serious dangers to life. Intellectuals, humanists, nonconformists, were despised in partisan units; and practices associated with civil society, for example, voting, had no place or context for practice. Those navigating the difficult ghetto escapes were young Jews capable of enduring physical hardship, finding weaponry, and surviving in harsh and minimalist conditions. The partisan units preferred strong leadership to consensus decision-making. In the forests, bickering could result in fatal delays. What Jews for centuries regarded as an uncivil way of life – the use of violence, extortion and murder, the submersion of individuality for group cohesion – transformed in the forests into the critical moral and political instruments of survival.

If one refused to adapt to the fighter's moral universe, to the concept of a hierarchy based on military prowess, then the likelihood of survival in the forest was greatly diminished. If one adapted to and accepted that reality, then survival was greatly enhanced. In the ghettos, suspicion of the underground, a refusal to accept its programs, repeated efforts to placate the Germans, the withdrawal into individualistic forms of adaptation, all these classic features of Jewish cultural life haunted the community until it was too late; until starvation, executions, mass reprisal and psychological desolation stifled any hope of collective resistance.

In *The Future of an Illusion*,[43] Freud argued that religion cripples and debilitates the self by making it dependent on illusions and not reality. But conditions in the forests proved Freud wrong. In the forest, the religion of vengeance, the communal values of group identity, so central to religious consciousness, were the major dynamic behind survival. Freud's scientific imagination, his rationalist bent, his skepticism, had little efficacy in the harsh life that faced those fortunate enough to escape the ghetto. The religion of the oppressed – vengeance – and the ideology of violence and retribution played a crucial role in cementing a generative relation between self and group and in assuring some chance of survival. Human and constitutional protections of dignity and life which we take so much for granted were quickly annihilated by German violence, and German barbarity succeeded in destroying political resistance in the ghettos. Those partisans who survived were resourceful enough and, in their words, 'lucky enough' to figure out ways to confront barbarism on its own terms, by the use of violence to cripple the enemy and break its will to dominate. But in my conversations with survivors, including those who were not part of any partisan or underground resistance, what appears with striking clarity is a theme returned to time and again: the belief that by killing Germans and collaborators, partisans were acting for all Jews.

5
Spiritual Resistance: Understanding its Meaning

Recording and witnessing: what is being seen

Diaries and survivor accounts are united by one overriding fact: the power of the German assault, its suddenness, its attack not only on body but also on mind and reason. Ghetto diaries describe human life slowly withering away; orphaned children, the elderly dying in the streets; food supplies diminishing; wealthy Jews turning in a matter of weeks into beggars; black markets and smuggling; families torn apart; people sinking into madness; random attacks from Germans and locals; and moral disintegration under the pressure of survival and the German policy of reprisals. Everywhere Jews looked, not only did brutality define the landscape but a community appeared whose leadership had fallen prey to German lies and whose spiritual leadership struggled with impossible human, moral and psychological demands.

Beaten, killed, starved, degraded, stripped of all possessions and dignity, the individual Jewish self undergoes trauma, madness, unimaginable from the vantage of normal day-to-day life. Madness comes to define much of daily life in the ghetto, and much in the rabbinical response to the Holocaust, including the writing and adapting of Jewish law, is an effort to contain, or at least provide some comfort to, a community facing madness and annihilation.

No area of Jewish life is more subject to ambiguity in understanding resistance, particularly regarding defenses against madness, than the role of spiritual and religious leaders *and* the place of theology during the Holocaust. Evidence appears in anecdotal form: rabbinical stories, first-person accounts, Jewish law or *Halakhah* written during the Holocaust, classical texts (Talmud, *Midrash, Kabbalah*); rabbinical recollections and reflection.[1] The evidence paints a picture of a theologically unshakeable and courageous rabbinate who, with few exceptions, opposed violent resistance.[2] Very little in the secular accounts (Zionist, communist, Bundist) questions rabbinical dignity.[3] Even partisan diaries are for the most part silent on the role of the rabbis. Some of the literature speaks of rabbis who supported underground activities, although such support was rare. The underground and partisan rage focuses on the Germans, the *Judenrate* and the Jewish police. These were the truly hated presence; but the resistance was focused. What Emmanuel Ringelblum calls the Jewish 'masses' were not. It is in this universe of rapidly declining moral expectations, the loss of hope and the rise of dissociated and mute selves, that theology and faith may have had some impact, or at least that's the argument I want to make.

While spiritual resistance may not have saved lives, it may have instilled a psychological refuge inside the self and within the community, where the self's identity as a person and a Jew could not be touched or degraded by German barbarism. This argument suggests that spiritual resistance and violent resistance reached for the same end: the preservation of self, or sanity, in a moral universe of rapid disintegration and loss of will. Spiritual resistance sought to save the soul, or psyche, from madness through silence in relation to the oppressor, but not silence in relation to God or to the Jewish community itself.

One of the strangest stories I heard concerning spiritual resistance was that of a Jewish partisan fighter, Sol, forced to masquerade as a Catholic. He did this, he told me, by 'always remembering that in my soul I was a Jew.' But the deception saved his life and his will: 'You ask me why it saved me? Because I got to kill Germans, blow up trains. Every time I shot a German, it felt good. That made it all worthwhile; their death meant nothing to me.' And throughout this time, all his outward religious practices were Catholic. 'I had to go to Mass. I had to do confession; but I'd do it all over again. Survive, survive, survive: that was what mattered. I had only one thing on my mind: survival. Sure I became religious during the war; religious like a Catholic. But inside, deep inside, I never forgot I was a Jew.'

However, for the vast majority of Jews, unlike Ben and the partisans, spiritual resistance possessed a singularly non-violent and silent component in the individual self's encounter with the aggressor. Many partisan survivors questioned the 'strategy' of silence; it did not save lives. But spiritual resistance took on political meaning and significance through the symbolism and action of refusal: refusal to succumb to the German assault on religious identity; refusal to allow the genocidal violence to kill the soul, and refusal to spiritually acquiesce to a system of values meant to annihilate the Jewish body and the Jewish faith. As an unknown diarist in the Lodz ghetto put it: 'Humanity is not dormant ... in its own manner your soul bears witness to the eternal, never-to-be vanquished victor who is called "human being".'[4]

Spiritual resistance *politically* confronted a lynchpin of German strategy: the demoralization of the Jewish community and their leadership, and the effort to destroy will, spirit and faith. Retaining faith in the historical community of the Jews, in the Talmud and cultural practice, was an important spiritual and psychological factor in efforts to counter the rising tide of despair that gripped the ghettos. It may be important here to distinguish between faith in God, a theological injunction, and faith in the community and its identity creating and sustaining functions. Partisan survivors continually made this distinction: to paraphrase, 'God, where was He?' as opposed to 'I never lost my faith in my Jewish identity, history, in the sacredness of practices, and my connections to the community itself ... God was somewhere, I guess, but I didn't know where.' Yet, given the increasing isolation of the ghettos and the muteness of the population, it is unclear whether faith mattered in preserving sanity, although I think sufficient evidence exists to suggest that it made some difference.

The example of the rabbis' courage, for example in attempting to protect Torah scrolls and objects of worship, or conducting religious services under trying circumstances, demonstrates a recognizable form of political resistance. The theological components of Judaism created a spiritual universe where acceptance of death with faith for many came to be regarded as a heroic response to German brutality. This acceptance was called Sanctification of the Name of the Lord. Martyrdom – *Kiddush haShem* – became for the rabbis, and for those retaining their faith until the end, the pivotal form of *communal* spiritual resistance. Theology had a profound effect on what the Jewish masses understood of fate and destiny. But it is not clear how strong a hold the *theological explanation of the Holocaust* had on the Jewish community or whether it provided a safe psychological harbor from German brutality. Certainly, many continued to believe in God, even in the undressing rooms of the gas chambers. But it is also true that psychological brutalization induced by the Germans had a profound impact on theological explanation. Religious faith and practice declined in the ghetto, partly because religious practices were banned by the Germans, but also because theology becomes progressively absent in daily life as conditions worsened. Religious faith amongst spiritual leaders becomes more intense, until death itself comes to be seen as a glorious sacrifice, *Kiddush haShem*.[5] For some, even in the grimmest moments of life in the ghetto, lighting a Sabbath candle provided relief from the horror of the outside. By the time individual Jews finally grasped the enormity of their condition, facing the entrance to the death chambers or standing beside the death pits, or watching friends and family die from disease, starvation or cold, the only spiritual universe remaining lay in the faith of martyrdom. How much that counted and for how many we will never know. The concept of spiritual resistance assumes that it may have counted for some.

Did individual Jews go to their death with the Lord's Name etched on their consciousness or in sheer bitterness and resignation or even already psychically dead? Evidence points to the fact that *Kiddush haShem* at the moment of death may have provided spiritual refuge and connection and helped individuals counter the panic accompanying imminent terror. While many survivor accounts reflect great disillusionment with theology and the religious establishment, it is wrong to impugn rabbinical commitment and the rabbinical struggle with theological interpretation. Rabbis demonstrated strong connection with their congregations. Even rabbis separated from their congregations performed religious ceremonies and led in prayer up to the very moment of death. Rabbinical action in protecting sacred objects constituted political resistance; it was public; it involved retaliation by the Germans; and the behavior itself symbolized limits that, if transgressed, would unleash retaliation from the enemy.[6] Not even the Polish army in the early days could halt the German advance, much less unarmed, provincial and politically unprepared Jewish spiritual leaders rushing into burning synagogues to save Torah scrolls.

Spiritual resistance provided an explanation and therefore a meaning for death.[7] Entering the gas chamber or standing at the edge of a pit waiting to be shot may not, then, have taken place in a spiritual and psychological void. With the presence and blessing of a rabbi, those waiting for death may have experienced refuge from the horror. No one can know what went through the minds of the millions who died. What we do know is that rabbis in the face of intolerable assault acted with great courage and spiritual resilience, and this alone may have been of critical importance in the bitter recognition of knowing that death was unavoidable. Spiritual resistance, by its very nature, whether in the ghettos or at the doors of the gas chambers, could not be expected to have had the same effect or consequence as violent resistance in rebuilding selves torn down by assault.[8]

Once in the forest or underground, the resistance fighter had freedom to act, to be, to define the world in terms unthinkable in the ghetto, where selves had been shattered by the trauma of assault. Yet, almost universally, survivors discounted spiritual resistance. It was a moral lapse in their view, a luxury without empirical consequence, with no impact on saving lives. Ben: 'That's absurd, to think that spiritual resistance could have been effective. Tell me, how do you transform spirit into bullets?' Or, as Miles Lerman put it: 'Where is spiritual resistance when you hang a collaborator and put a sign on his body saying "This is what happens to collaborators." We killed and refused to kneel. We were prepared to die... . I remember Moishe, a *cheder* boy; but was he fierce! In a shoot-out he choked a guy to death, grabbed his machine gun and mowed down some four or five guys. It wasn't prayer that did that.'

It is true that in the ghettos religious practice declined, given the demands of basic day-to-day survival and German proscription of religious practice. But religious identity never disappeared. The Jews remained acutely aware of their identity as Jews, not as a negative projection, but as a culture singled out for terrible punishment.[9] The rabbis and the theology attempted to counter the negative German projection by clinging to biblical and Talmudic interpretation to reinforce Jewish cultural identity, to resist German assault on beliefs central to the meaning of being a Jew. The rabbis provided a vision of the Jewish self in which suffering possessed a divine significance.[10] While from a contemporary perspective this may not be satisfactory as resistance, it would be wrong to take away from the theology this critically important spiritual function in a time of frightening uncertainty.[11]

Rabbinical writings and sermons address the issue of theological identity, sometimes with reference to the Holocaust, but often as a commentary on scholarly and Talmudic analysis reflecting on evil, the destruction of the Jewish people and the assault on Jewish identity, practice and property. While religious expression of faith in God diminished in the ghettos, religious *affirmation* continued as text and exegesis in rabbinical writings, judgments and law for guidance in morally problematic, human dilemmas. For example, a Polish man offered a Jewish girl safety and protection in exchange for sleeping with him. She asks for the rabbi's guidance. The rabbi formulated a 'law' which allowed her to retain her self-respect, yet at the same time take advantage of an opportunity to escape near-certain death. He granted her what amounted to a Talmudic dispensation, acknowledging that strict moral interpretation of law must be balanced against the severity of the times and the sanctity of human life.

Every day, rabbis had to deal with such dilemmas, what Primo Levi calls the 'gray zone' in human and spiritual behavior.[12] But underlying these moral quandaries lay a deep despair, a recognition of the power of the German death machine. Shalom Cholawski recalls: 'I will always remember the sight of the mother as she watched her children being dragged away by the Germans. She was hitting her head against the wall, as if to punish herself for remaining silent, for wanting to live.'[13]

Rabbinical authority, with a few exceptions, never called for violent, armed resistance; rabbis rarely came into conflict with the *Judenrate*. *Judenrat* leaders requested guidance from rabbinical councils on how to handle German demands. Talmudic law and texts would be consulted and then interpretations given: Should the *Judenrat* hand out labor cards exempting some but not all from selection? Should the *Judenrat* comply with German demands to assemble 20,000 people in the town square, enabling the SS to make selections for the death camps? Political questions, requiring theological interpretation, involved how far to comply with German demands in deciding who was to live and die, how food should be distributed, where and how apartments and labor tasks should be allocated. As the terror and starvation increased in the ghettos and the random actions of brutality intensified, the rabbis focused increasingly on theological justifications for unprecedented suffering.[14]

Little communication, particularly in most of the larger ghettos, was established between rabbinical councils and resistance groups. Rabbis themselves saw their role in the resistance as fundamentally spiritual, behaving with dignity in the face of German assaults, protecting religious objects and Torah scrolls, comforting congregants and easing the spiritual path to death. The question is not whether more could have been done, but how the Germans were able to construct a universe of terror which assured the impossibility of doing more. Rabbinical authority sought to counter the power of the German psychological and physical advance; ultimately, though, the onslaught devoured resistance and made it impossible, with a few notable exceptions, to organize resistance in the ghettos themselves. For the rabbis, consulting the Talmud and protecting the Torah became the primary form of *action*, but against the *Einsatzgruppen*, Zyklon B gas, the slave labor camps, medical experiments, mass slaughter in the forest, starvation and disease, Talmudic law and God's Word could not be transformed into action. In Lerman's words: 'We knew we were Jews; we observed sacred days like Yom Kippur and Passover as much as we could. But we had no time to think about theology.'

The Jewish community invested enormously in the moral authority of the rabbis, and rabbinical judgment stood guard over the community's belief structures, its identity in times of peril, and the foundation of its ethical systems. The spiritual authority of the rabbis literally directed the community's moral inventory. *Judenrat* leaders during the Holocaust carried out administrative and political functions; the moral imperatives and justifications underlying the community's spiritual welfare and action lay in the hands of theological authority. But as Lerman puts it, 'God couldn't pull a trigger.' For resistance survivors protecting the soul before God should not be considered resistance, but a moral reckoning at the moment of death.

By spring 1943, when it appeared that God's rescue was not imminent, the moral authority of the rabbis increasingly turned to justifying theology, taking on greater urgency as the community found itself unable to temper the German assault. A question asked by a young man of his rabbi suggests how desperate people had become in the search for explanation and meaning: 'Rabbi, why doesn't the Messiah come? These are such terrible times. How much worse do things have to become before the promise of Redemption is fulfilled? If there is not the End of Days, then there is no End! Rabbi, I demand a hearing with the Messiah.'[15]

Compelling evidence, particularly in contemporaneous diary accounts, however, suggests that psychological processes far graver than a belief in martyrdom had surfaced in the lives of those in the ghetto. At the same time, the community appeared to be looking for explanations for the horror of its situation. Hasidic belief in the Messiah is a real and vital part of theology. In the absence of the Messiah's coming, however, martyrdom offered theological hope, but also reinforced quiescence in the community regarding the organization of violent action. *Realpolitik* had no presence in theology.[16] One resistance survivor told me: 'My parents, I remember, refused to protect themselves, trusting that God would save them. I pleaded with them to escape, to run to the forests, anything. But they just stayed.'

Spirit and survival: the protections of the inner self

Vernon describes how he survived Auschwitz: 'I was not going to let the Germans steal my soul; I decided that from the first day. It was not a matter of belief in God. For me that question had no meaning; and it was useless to talk about it. Sometimes guys would cry out against God; others would just be silent. But I knew that if I was going to survive, I had to hold to God because that was the only way I could hold on to my soul.' At this point in our conversation, after about two hours, Vernon leans over and taps my knee and, in an almost conspiratorial voice, says: 'Listen, you really want to know about spiritual resistance; I'll tell you. I've never told anyone this, not even my wife; but I kept my soul from the Germans by praying.' God becomes a living presence, not as an actor or non-actor in the camp, but as a presence in time and memory. That was enough for Vernon.

> 'I understood that God could not control every human being; that he had no power to choose who was to live and who to die.

Once I decided that, it was no longer "Is God there or not?" But "Where can I find God?" And I found Him in thinking about who I was, where I had been before Auschwitz. God had no power over the Germans and their brutality. But if I could have a place inside myself that the Germans couldn't touch, maybe that's where God could be in a place like Auschwitz. That was my thinking: the Germans had no godliness, so how could God have anything to do with what the Germans were doing?'

Vernon works this out as an operative theology to protect the boundaries of his soul; these thoughts wrap his 'soul' in what he calls 'a steel fence keeping the Germans out.' To maintain that fence, he had to think about God, keep God in his consciousness day in and day out:

'So I said to myself – I never had the strength to talk about this with any of the other prisoners – God is there but in a different shape, a form you can't see, but He lives inside of us and outside of us. He surrounds and whispers to those heading for the gas chambers. And no matter what happened I refused to give up that belief.'

Vernon accepted the fact that God could not control Auschwitz or the murder; but refused to give up his belief that God was 'out there.' It wasn't that God was hiding, but that in the spheres of responsibility that prevail in the world of spirit and secular power God could act only in the world of spirit. 'I convinced myself that God had given each of us a responsibility, our own godliness. And what we had to do was maintain that godliness. I did it through prayer.' Prayers, daily thought and uttered, changed and transformed, reinforced the steel fence surrounding his soul; prayer kept the Germans out.

'Every day I prayed to God, not to save me, but to have a presence that I knew was there, so I could do things that would make me worthy of godliness – godliness in God's eyes. Not in the eyes of those I knew in the camp or the Germans, but only in my eyes and God's. I could be worthy, I decided, by remaining alive; that would be the sign of my godliness. So I made up prayers; I would add to the prayers, add a phrase in Hebrew, change it. Praying helped me to see myself not as some ruined human being, some Mussulman wretch, but as someone who partakes of godliness even in a place as horrible as Auschwitz. Each day I survived, I knew it had been because of my prayer and that God by making me godly had not disappeared. The sign of my godliness was in my surviving each day.'

He used this belief to distance himself psychologically from the Germans. 'I got strength from knowing that our German tormentors had no godliness; so if they had no godliness, it was impossible for them to be part of what is good. God could only be with those who have godliness, so I could not blame Him for what the Germans did to us.' He admits this connection with faith came not from study but from his own psychological inventiveness; he is convinced that without this inner space of freedom, he would have died.

'I don't know if what I thought was consistent with Torah or not, but it kept my spirit up; it kept me alive. You see the question for all of us was to deal with this horrible place: how do you keep the Germans from trampling your own soul? You needed to figure something out; and prayer was one way I did it. It gave me a sense of who I was, apart from being someone being degraded, beaten and starved. By praying I knew I was a Jew; because I could speak the prayer, to myself silently, the very utterance of the words meant I was alive and surviving; if I could pray, I was alive; if I couldn't pray, I would be dead. Prayer didn't take me out of misery; don't misunderstand me; I suffered there. But prayer kept me from sinking into apathy and letting the camp kill me.'

Spirit as a form of prayer became a critical means of Vernon's psychological survival; yet, this sentiment rarely appeared in partisan survivor accounts. More typical of partisan survivors are the bitter words of Ben, who fought with a Soviet unit. I asked him about spiritual resistance and whether belief or faith in God mattered. His response: 'When you are stinking and weak from hunger, with lice crawling all over your body, spiritual things disappear; you only think about, look at, what's in front of your face. I was too busy to think about God.' When I asked him if there were any particular instances that he could remember which directly affected his faith in God's presence, he looked at me, with tears streaming down his face. 'When you see a Polish man cut off the breast of a Jewish woman, you ask, where is God? ... The life we had to go through, religion had no place in it. It was gone, for a long time.' But then he acknowledged: 'Many Jews prayed in the forest; I couldn't even though I was brought up in an orthodox home. What kept the Germans out was a bullet, not prayers.'

Where survivors like Vernon and Ben share common ground is in the spiritual sustenance of customary practice and the knowledge that no German could take away Jewish identity and history. For Ben lighting a *shabbes* candle had as much significance as prayer did for Vernon. In this sense spirit mattered, but a spirit rooted in Jewish memory and the words of the law, in the pride of surviving the German assault against that identity.

It is unrealistic to demand violent resistance from a spiritual universe that draws its explanatory propositions from the world of ancient theological texts. Perhaps this explains some of the absence of underground and partisan criticism of rabbinical authority.

A rabbi is being humiliated by the Germans before his congregation; he speaks to them while being beaten: 'Jews, you were mistaken about me all these years. I was never a righteous man, I wasn't even really pious. I have been a sinner all my life. God is right in bringing this terrible punishment.'[17] Now, one can admire the rabbi's faith; but it is also the case that for many Jews belief in God lost all meaning. I showed this passage to one survivor, who remarked, 'That rabbi was a fool.'

For resistance survivors, the group or politics replaced theology; group consciousness became the container and the spiritual point of reference. Indeed, the group in Rousseauian terms possesses a general will: defined as the political ethics of revenge and rescue. Lerman puts it this way:

'The relationships with the group were very close; there was much love, but discipline too. Everyone had to do their part; there were women and some older children, around twelve or thirteen. We sent the girls into town, but not the boys. Pull down a boy's pants and you immediately know he is a Jew. The group made us feel human; it helped our despair and the hopelessness of knowing everyone you had loved was dead. Each of us felt we made some difference; that life had purpose and meaning – and that purpose had to do with saving ourselves and with taking revenge. In those forests, we were a force to be reckoned with; we were not passive like dogs. We killed, and that made us strong. It is just not possible to imagine what the Germans did to us; to kill was not just revenge; it was coming back into the world of the living ...'

For Lerman to be alive meant to kill, not as an isolated murderer or soldier, but as a political extension of a group will defining itself in the new, forest-based terms of Jewish identity:

'For me to have any psychological health I had to know, had to see, the enemy *dead*. We bothered the Germans so much they put posters on the trees to warn the local populations against the bandits. So, we put our own posters on the trees. They said: "The bandits were here." ... I remember one night, it was after a fight with the Ukrainians; we lost a few.'

Arguing with God: a dialogue regarding faith

A very different and non-political approach to spiritual resistance appears in the homilies and sermons of Rabbi Kalonymus Kalman Shapira, the leader of a small congregation in Warsaw. Shapira's homilies, written between 1939 and 1942, constitute a remarkable exegesis of textual matters pertaining to faith. Rabbi Shapira died in November 1943 in a labor camp near Lublin, but his writings, collected in a volume of reflections called *The Holy Fire*, testify to the spiritual resilience of faith and courage in view of his terrible personal tragedy. Shapira's entire family – his wife and other family members – were waiting outside a Warsaw hospital for news of his seriously ill son when they were killed by a bomb blast.

Shapira's sermons demonstrate the rigors of spiritual resistance and reveal the psychological and spiritual struggle Jews faced when the Germans dismantled the ghetto and transported hundreds of thousands of Jews to Sobibor and Treblinka. He articulates a spiritual/theological position which has nothing in common with the undergrounds or partisans, but his words constitute a profound negation of spiritual quietism, coupled with an absolute unwillingness to question God's intent. Even more important, at least for our purposes, the homilies give us an insight into how fragile spiritual values and resistance became in the final days of the Warsaw ghetto.

Rabbi Shapira argued with God about German desecration. 'How can you tolerate the humiliation of the Torah, and Israel's anguish? They [those in the ghetto] are being tormented and tortured just because they fulfill the Torah.'[18] This bitter, private conversation is consistent with the rabbinical tradition of arguing with God. When questions arise outside a relationship to God, then dispute with God is impermissible and constitutes a violation of faith. The legitimacy of questioning God depends on where the 'self' stands. Rabbi Shapira shares his anguish with God: 'When we hear the voices of young and old crying out under torture, "*Ratevet! Ratevet!*" [Help! Help!], we know that this is their souls' cry, and the cry of all our souls to God, the compassionate Father – Help! Help!, while the breath of life is still within us.'[19] Why does God not listen? 'But now innocent children, pure angels, as well as adults, the saintly of Israel, are killed and slaughtered just because they are Jews, who are greater than angels. They fill the entire space of the universe with these cries, and the world does not turn back to water, but remains in place.'[20] God, however, suffers for Israel. He weeps, but not in the outer world; his weeping is invisible.

When I quoted this passage to Ben, he said: 'What use are invisible tears? So what if God is crying privately? It certainly didn't help anyone.' Shapira: 'In His inner chambers [God] grieves and weeps for the sufferings of Israel ... one who pushes in and comes close to Him by means of studying Torah weeps together with God, and studies Torah with Him.'[21] God's concealment derives not from His indifference to Israel but because He is hidden in His chambers, weeping for Israel. God suffers too because in attacking Israel and the Jewish people, the Germans also attack God; their hatred 'is basically for the Torah, and as a consequence they torment us as well.'[22]

But arguing with God meant little in the context of German policies. As one survivor put it to me: 'What could God do to destroy the gas chambers ... "arguing with God" – that was crazy. The Germans wanted us dead. It was as simple as that.'

Yet, in the midst of evidence demonstrating conclusively genocidal action, Rabbi Shapira insisted that God could see the torment Jews suffer. (The resistance fighter's response to this position was 'If God could see, why did he act so blindly?') He is witness to the evil. However, the evil of *Din* [a force of great severity and destructiveness], its murderousness and ferocity, its presence in the world as a 'torturing and tormenting ... serpent ... a carnivorous, devouring beast' is to be vanquished by 'a supernatural effort.' Ben laughed at this passage: 'Supernatural effort? That was us; we did supernatural efforts in fighting the Germans.' But Shapira believed that at some point in the future evil would be transformed into good; that even in the midst of a horrifying reality lies an inner 'light,' which, although hidden, requires the right condition for it to be 'revealed so that everything will be transmuted, or sweetened into *Rachamin* [pity, compassion].'[23]

One can only speculate on what was required of the imagination to accept these teachings in 1942, for the congregants to believe that when one listens to the voice of the Torah, one hears it in 'the chirping of the birds, the mooing of the cows, the voices and tumult of human beings.'[24] Or that through the action of prayer 'all evil is sublimated into good; all evil utterance, all evil discourse which Israel's enemies utter against her, is all transformed into the voice of Torah.' Those 'conditions' denying evil will be transmuted into 'sweetness.'[25] But nothing of that 'sweetness' is found in the boxcars destined for Treblinka and Sobibor or in the screams of children watching their parents beaten and killed. Ben: 'Sweetness? Is this guy crazy or what? ... sweetness when you watch your children die? I could never accept an argument like that!'

We have no quantifiable, objective criteria for assessing how Rabbi Shapira's theology was received in the Warsaw of spring/summer 1942, when hundreds of thousands of Jews were being transported to the death camps, but from the diaries and eyewitness reports, it is likely that the thousands of Jews herded into the central plaza of the ghetto to await transport experienced numbness and disbelief rather than the consolations of faith.[26]

Dissociative psychological process: not seeing the pain

Genocide requires of the executioner psychological factors that destroy empathy, allowing the killers to proceed with no moral qualms. Dissociation radically distorts perception and may have a tremendous impact on how a person or object is *experienced*. The dissociative response acts as a barrier or shield to the sensibility, presence and suffering of the Other. I recall a patient, Julia [a pseudonym], during my research at the Shepard and Enoch Pratt Hospital in Towson, Maryland, who told me she once tried to win her mother's attention by carving into her thigh, with a sharp kitchen knife, the words 'I love you, Mom.' Dripping with blood, she walked into her mother's dressing room where her mother was sitting in front of a large mirror applying make-up. In the mirror's reflection she could clearly see her daughter; but did not acknowledge her daughter's bleeding leg. Julia's mother literally had dissociated her daughter's suffering from her own perceptual range; she experienced Julia as interrupting her make-up ritual, interfering with her application of mascara. It is only after she noticed blood on the rug that she finally *saw* her daughter's wound. But by then too much psychological damage had been done; weeping, Julia had run out of the room.

What does it take for anyone *not* to see or feel the suffering or presence of others? That becomes a particularly compelling question in the context of suffering and brutalization in the ghettos. Ben suggested that the Germans and God might have had something in common: neither could really feel, much less empathize with, the suffering of the Jews. Many Germans saw the suffering; many knew what was going on, if not those who had business in the ghettos, then employers who used slave labor or German workers employed in the same factories, or German housewives buying Jewish clothing, goods and artifacts at auctions held throughout the country, or in any of the hundreds of slave labor camps inside Germany which required periodic logistical contact with local populations. The evidence was everywhere: emaciated bodies, tattered uniforms, vacant eyes, constantly appeared in ordinary life in Germany. During the death marches thousands of starved and dying bodies dragged themselves through German towns, only to be met with jeers, derision, brutality and sadistic jests. Why? What forges barriers to acknowledging and experiencing the Other's psychological suffering?

The German executioners had transformed the Jewish body into a feared and hated object. *Not* to see the suffering imposed by starvation, *not* to see pain in children's faces as they were torn from their mothers, *not* to see anything wrong in organizing kindergartens in ghettos for the purpose of concentrating Jewish children for seizure and transport to death camps, *not* to feel remorse for executing entire families in open pits – all this suggests a moral environment legitimating and encouraging the disposal of what is perceived as diseased flesh. Not to see a five-year-old child pleading for his life, but to see an enemy to one's biological existence accounts for the moral ease that accompanied the killing process in and around the ghettos. With those psychological processes conditioning perception in the oppressor, Rabbi Shapira's faith operated in an emotional vacuum; it defined only an internal psychological space in the victim and it allowed the self literally to curl up within itself and turn away from reality. One can admire his conversations with God. But resistance survivors expressed little, if any, compassion for his theological stance.

Dissociation of the Other facilitates radical transformations in how environments are perceived. Christopher Browning, Daniel Goldhagen, Gotz Aly and Benno Muller Hill[27] have all documented the ease with which individuals moved in and out of killing environments without moral qualms or guilt. On rare occasions when individual Germans assigned to killing squads protested the slaughter or found it intolerable to participate, reassignment to another job or unit quickly followed. No German was executed because he or she refused to murder Jews.

To see the Other not as a victim, not as a sufferer, but as threat to the boundaries of self and culture placed the executioner in a unique relationship with the victim. Rabbi Shapira's theology possessed no moral category to account for this. The Germans' psychological and biological fear of the Jews had no role in the dialogue with God, because God and Rabbi Shapira's interpretation of God could not comprehend how the German vision of biological health defined 'victimhood' as toxic waste, infected Jewish bodies to be burned. As Simon put it: 'If God could have known how frightened the Germans were of Jewish bodies, would He have hidden His tears?' For the Germans, what was being killed were not people, but fears and hatreds displaced onto innocent bodies. In killing Jews, the Germans were maintaining not only the highest standard of what they regarded as effective and willing citizenship, but also the 'noble' purposes of a moral order assuring, forever, blood purity and racial integrity. It was in the name of these biological ideals and the fear of being tainted by diseased flesh that a moral environment was legitimated which had as its singular purpose the annihilation of Jewish bodies.

Yet, even though Rabbi Shapira's dialogue took place in a moral vacuum, without political content, it needs to be stressed that his homilies contained affirmations desperately attempting to maintain contact with a shattered historical and cultural reality. His reaction, then, to the oppressor's dissociative actions took on essentially two forms: an assertion of the sovereignty of God's Word and pleas not to forsake God; and an argument for an emotional or irrational fusion with faith, an unquestioning acceptance of God's presence. But whether it was faith in God, faith in Jewish identity and culture, or faith in the power of witnessing and remembrance, it would be wrong to discount this form of psychological positioning towards the oppressor. We can never know how many retained spiritual protest in their consciousness; but it may have been significant. No indication appears in the *d'rashot* [the theological teachings] of whether Rabbi Shapira sanctioned or condemned armed resistance. But his silence on the resistance in Warsaw indicates at least a passive acceptance of its utility. What we do know is that Rabbi Shapira's own resistance took place entirely within a theological belief system, an absolute trust in God, and an unquestioning acceptance of God's will. Rabbi Shapira urged his congregants to respond to the sufferings of the community, not to close off their hearts, to remain in contact with God and with communal efforts to alleviate the suffering. *Emunah,* or faith, indicates the presence of an active Jewish self, communicating, always aware of the boundaries of its identity; the self stays alive only if it affirms devotion to God. Even in spite of its political powerlessness, this faith may have made some difference to those who suffered.

6
Condemned Spirit and the Moral Arguments of Faith

Introduction

In his treatise on resurrection, Moses Maimonides writes: 'the resurrection of the dead is one of the cornerstones of the Torah ... there is no portion for him that denies that it is part of the Torah of Moses our Teacher.'[1] Having established the theological legitimacy of resurrection, Maimonides goes on to argue that while resurrection is critical as a theological component of Judaism, it is not the 'ultimate goal' of religious faith. For the believer there is no higher hope than the world to come; nothing in faith achieves the 'bliss beyond which there is none more blissful' than in the world to come. 'Corporeal existence' disappears since God is 'not corporeal nor a power within a body and, therefore, the level of His existence is the firmest of all.'[2]

Belief in a higher world of non-corporeal bodies became a central tenet of faith during the Holocaust – a faith that acted as a protest against the German vision of wiping out 'life unworthy of life' from all existence. What faith, as protest, meant in this context lay in the vision of moving not only to another level or place of 'existence,' but the possibility of redemption beyond the suffering of this world. Rabbis taught through this period that death could be understood not as nothingness, but as a moment of the supreme testing of faith. These beliefs were being articulated by rabbis in the ghettos of the East, to make sense of the utter confusion inflicted by the German assault.

For Maimonides, Moses' injunction was absolute: '*Hear, O Israel, the Lord is our God, the Lord is one.*'[3] Rabbi Shapira took this literally:

obedience and faith to the God of Israel could not be questioned; to doubt faith meant giving a victory to the Germans. While abuse could be relentless, faith constituted an untouchable place within the self. God defined all destiny: the body might suffer brutality, but faith would endure. An ancient Biblical prediction that the Messiah would come in the year 5700 (1940) kept many Jews within the orbit of faith through 1940–41. Signs and their interpretation of the impending coming of the Messiah turned into heated arguments. One Hasidic rabbi preached: 'When we begin to blow the *shofar* [ram's horn], the enemy will be blown away.'[4] A number of signs were found in religious texts, in phrases or placement of letters, pagination. People searched desperately for any textual or experiential event out of the ordinary for clues as to where and when the Messiah would come.

Early in the occupation of Poland, a few rabbis responded violently to Jewish collaboration in the German occupation; a rabbi ordered an informer's tongue to be cut out. In the Lublin area Jews believed to be informers were drowned in *mikvah* baths.[5] Ritual practice would take on the form of resistance; for example, walking to the *mikvah* bath or the ritual slaughterhouses would be a cause for arrest, torture and execution.[6] Warnings had been posted throughout the ghetto: 'Opening the *mikveh* or employing it will be punished as sabotage and will be subject to between ten years in prison and death.'[7] Religious practice, then, could be understood as resistance; whether one fired a gun or immersed the body in a sacred bath, the consequences could be the same.

Children participated in ritual resistance by smuggling kosher-slaughtered meat into the ghettos; they found breaches in ghetto walls, sneaked through entry points or arranged with confederates to receive merchandise. On occasion, children as young as five or six would hide contraband on their bodies, taking great risks at the ghetto gates. Ritual slaughterers themselves incurred risk, since all kosher practices had been banned by the Germans. To be caught in these acts meant certain death. Kosher meat early in the occupation found its way into the ghetto through a variety of conveyances – in trucks belonging to the utilities companies, medical resources, services, and municipal sanitation vehicles. Any driver, Pole or Jew, caught with kosher meat was immediately arrested.[8]

Rabbi Shapira's call to faith occurs in a universe filled with death and random executions. An underground Warsaw newspaper has an article entitled, 'A Dance amidst Corpses'; the author writes:

> 'The situation in the Jewish ghetto is fatal. On every street corner lies the corpse of a Jew who has died of hunger and cold. Nearly 7,000 funerals take place each month. And at the very same time, the Jewish cabarets and nightclubs conduct dancing contests with enormous prizes. In these nightclubs, tens of thousands of zlotys are wasted each night. This is truly a dance in the midst of corpses.'[9]

By the spring of 1942, class distinctions no longer existed in the ghetto; all Jewish life had been reduced to poverty, starvation and death. The rich, at least initially, turned a blind eye to suffering; but the evidence suggests that money made a difference for a few months at most. More common, however, than accounts of the indifference of rich Jews are witness recollections of protests taking the form of protection of sacred texts. In a small village called Sierpe, the Germans ordered the Jewish population to watch the burning down of a synagogue. Shortly after the *shul* was set ablaze, a young man dashed into the burning building; then reappeared with the Torah scrolls in his arms. The Germans shot him and his body, still clutching the scrolls, was engulfed by flames. It was problematic if one could retain sanity and psychic balance amidst such horror. Consider the matter of funerals, a practice or rite we take for granted. Not for the ghettoized Jews; one could not even be sure a body would be delivered to the cemetery:

> 'Today is already my second day of waiting here at the cemetery. I wait and wait, and my daughter hasn't even been brought to the cemetery yet. I don't know where to look for her. They told me in the office that a wagon of corpses should be arriving soon from Warsaw. Maybe my child will finally be on that transport. I don't know what has become of my dear child. Such a precious child – so clever – so beautiful – so many wonderful traits. My poor baby fell ill because of the contaminated water we drank. She was sick for a few months and then – woe is me – she was gone.'[10]

Rabbis faced unprecedented moral and ethical dilemmas unique in the life of a culture whose very existence was being eradicated. The head of a distinguished family in Kovno asked a rabbi if it would be religiously permissible for him to commit suicide. Rabbi Oshry writes:

> 'Is it permissible to hasten his end, to set his own hand against himself, even though it not be lawful? This, so that his own eyes do not see the destruction of his family. And so that he himself does not die a violent death after unspeakable tortures by the "cursed murderers, may their name be blotted out"; and so that he might have the merit of being given a proper Jewish burial in the Jewish cemetery of the Kovno ghetto.'[11]

While suicide is prohibited, and in the *Midrash* is considered a 'grave sin,' when suicide is committed for the 'sanctification of the Lord,' it is not an act to be morally condemned. Suicides, which occurred on a daily basis in many ghettos, Rabbi Oshry concludes, under the kind of duress imposed on the Jews by the Germans, cannot be considered to preclude customary mourning practices by relatives.[12]

A Hasidic Jew in a small Polish ghetto describes German atrocities:

> 'What can they do to me? They can take my body – but not my soul! Over my soul they have no dominion! Their dominion is only in this world. Here they are the mighty ones. All right. But in the world to come their strength is no more.'[13]

Again, for many Jews this protection of the inner space of faith, an unwillingness to let it be broken, became the only form of resistance.

Physical assault and disorientation

Starvation, exposure, beating; rounding up for transport to unfamiliar ghettos; dislocation provoked by transport itself; the disappearance of friends and relatives, confiscation of property and money; relocation of families from villages and towns to ghettos established specifically for the purpose of concentrating Jewish populations – all were German strategies for debilitating the will and resistance. In

Krakow, for example, Jews who had lived in Kazimierz for centuries were moved to Podgoreze, a run-down district on the outskirts of the city. Several families had to share one room; 20,000 people occupied a walled-in area of around 320 buildings, whose length could be walked, at a normal pace, in seven to ten minutes. The dislocation following uprooting made control easier for the Germans; in segregating the Jewish population, German administrators enforced dependency on the victimizer, severed the connection of the self to day-to-day productivity and habit and weakened the self's ability to navigate unfamiliar environments. Adaptation to debilitating, unfamiliar physical conditions, to the habits of strangers, to extensive food shortages, to poor sanitation, burdened individuals with the sheer minute-to-minute demands of survival. Yet many believed that if they could sustain and demonstrate their faith, God might intervene. The other side of it was put to me by Aaron (Bielski) Bell: 'Sure, you could have faith; but by the time God came, you'd be dead. And what's the use of faith if you're dead.'

In the Krakow city archives, German health documents describe health practices during this period. These extraordinary letters, memos and lists demonstrate how insistent the Germans were in strictly segregating the Jewish and Aryan populations because of the fear of infection, particularly the belief that Jews, because of their biology and blood type, were innate carriers of typhus. But Jews had little understanding of how powerful this phobia was in determining German intent and practices regarding 'infected' bodies. In a memo prepared by the German 'Disinfection' Chief, Dr. Kroll, a justification for the establishment of the Krakow ghetto was the belief that it would prevent the spread of disease between Jewish and Polish populations; streetcars or trams passing through the ghetto strictly enforced the policy of no contact between Pole and Jew. In Warsaw, trams traversing the ghetto had their windows boarded up.

Onerous demands were placed on Jewish doctors, dentists and professionals during the early days of the Polish occupation. Jews paid higher taxes and license fees than Poles, while the practices of Jewish professionals were progressively restricted. Jewish doctors could not participate in the Polish health insurance system. Detailed racial inventories were kept of health practitioners, especially doctors and dentists. Jews filled out complex forms describing racial history, paid additional taxes on everything from supplies to income to living and workplaces. Jewish doctors were not allowed to treat non-Jewish patients; they could be put to death for touching non-Jews – and all this as early as winter 1940. Thousands of petitions written by Jewish professionals protesting the financial burdens and practice restrictions were registered and denied. Polish physicians, however, were allowed to practice, as long as they treated only Polish patients. The Germans believed the Poles would be the workers, the muscles of the coming Reich millennium; Polish workers needed food and required medical supplies and treatment. Many Polish intellectuals, however, especially those in universities in Krakow and Warsaw, perished in Auschwitz.

The Krakow health administration archives present a striking picture of how obsessed German public health officials were with Jewish diseases and bodies and the extent of bureaucratic participation in maintaining strict segregation and inequality. For example, prescriptions filled by Jewish pharmacists had to be stamped with the Star of David. Rerouted trolley lines avoided contact with Jewish ghettos. In a memo to the head of the Polish Chamber of Physicians, the chief German Sanitation Inspector, Dr. Kroll, argued one of the benefits of establishing the Krakow ghetto in Podgoreze lay in improving traffic patterns and isolating disease. All aspects of public health administration – which meant isolating, containing and killing Jews – fell under the auspices of German administrators. Doctors brought in from the Ukraine supervised Polish physicians in hospitals, since the Germans believed Polish doctors might treat wounded resistance fighters. German doctors from the Reich tended Germans in occupied Poland.

Nadia S., in a phone interview, recalls this period: 'You can't imagine how bad it was; Germans, Poles, they spat on us; we had a nice apartment in Krakow but we were moved into one room with three other families. Germans would shoot us, often like a game. It was hard to think of God in that ghetto. Sure, I had faith as a Jew, but prayer? There was too much suffering even to think about praying.'

The effects of the race laws could be seen where food was distributed (spoiled rations, insufficient supplies, outrageous prices); workshops (terrible working conditions); soup kitchens (woefully inadequate nourishment); orphanages (thousands of homeless, abandoned children); shelter (terrible sanitation, little or no bedding, almost no fuel to protect against the harsh winters); medical treatment deficiencies;

understaffed and filthy hospitals; four or more families living in one room; holes in the ground serving as toilets. At every social, cultural, economic, political and moral site, Jews not only experienced brutality in food distribution, medical services and wages, administrative functions and priorities, and the elemental decency that makes life possible, but constantly faced the annihilatory logic and terror of genocide. In the German-sponsored Warsaw daily newspaper, reports periodically appeared noting 'de-Jewed' industries, which meant that the Jewish proprietors had been murdered outright, transported to a ghetto or shipped to a death camp. Rabbis advocating 'faith' lived in the midst of these atrocities and the ideology, authorized by German science, that Jews possessed innate biological and genetic dangers to Germans. Theology simply had no way of comprehending how absolute this vision was and how determined scientific authority was in eradicating this biological/blood threat.

Janusz Korczak captures something of the confusion brought on by radical dislocation and the Germans' utter disregard for the fate of Jewish bodies. 'What matters is that all this did happen ... the destitute beggars suspended between prison and hospital. The slave work ... debased faith, family, motherhood.'[14] Korczak, an old man by this time, walked to the *Umschlagplatz* [collection point for transport to Sobibor and Treblinka] with the two hundred children in his Orphanage, even though the Germans offered to spare his life. Leading that slow walk to the place of deportation, children clinging to him, calmed by his attention and care, Korczak embodied a spiritual presence not even the Germans could destroy. Korczak's action said: 'You will not destroy my dignity, my humanity.' Korczak was immensely popular not only with assimilated, urban Jews but with the Polish population; he had been the equivalent of Dr. Benjamin Spock before the war; his radio program on pediatric care reached the entire Polish population. By rejecting the offer to spare his life, Korczak escaped a central element of German policy towards the Jewish population: debasing the spirit and destroying the will even before killing the body.

The ghetto site embodied the extreme horror of the attack on Poland's Jews. For example, Korczak on Warsaw:

'The look of this district is changing from day to day. 1. A prison. 2. A plague-stricken area. 3. A mating ground. 4. A lunatic asylum. 5. A casino, Monaco. The stake – your head … . A young boy, still alive or perhaps dead already, is lying across the sidewalk. Right there three boys are playing horses and drivers; their reins have gotten entangled. They try every which way to disentangle them, they grow impatient, stumble over the boy lying on the ground. Finally one of them says, "Let's move on, he gets in the way." They move a few steps away and continue to struggle with the reins.'[15]

Rabbi Shapira was writing in this insane Warsaw universe from 1939 to 1942; he called his sermons *Hiddushei Torah auf Sedros* [New Torah Insights]. The last entry is dated July 18, 1942, less than a week before mass deportations from Warsaw to Treblinka. Shapira never questions his faith, even in the midst of intense personal suffering; and his theological writings are filled with references from the Talmud and *Midrash*. He refuses to utter the word German or Nazi; nor does he address the politics of his times or the desperate plight of refugees streaming into Warsaw or the collaborationist policies of the *Judenrat*. Even facing his own suffering after the loss of his family, Rabbi Shapira stays within the boundaries of the theological text. He sermonizes on spiritual malaise, and the *Esh Kodesh* [Holy Fire] involves an ongoing spiritual dialogue between God and himself. Shapira responds to the weakening of religious observance, the turning away from faith and the lack of enthusiasm for religious explanation, with the only 'weapon' he knows: the power of theological text.

God's presence in the self

Shapira struggles with his torment regarding the constancy of God's presence: 'There are times,' he writes, when God might 'smite us,' thereby creating a 'distance' between 'us' and 'Him.' But even in approaching the threshold of doubt, Shapira turns away and affirms God's truth. It is not for the individual to 'say if something is a plague or calamity,' he might say it 'seems' to be a calamity. 'The truth' is that whatever God intends, no matter how disastrous, 'is a good for Israel; God will bestow good upon us.'[16]

Even as the German assault pushes the ghetto further from religious consciousness, Shapira develops theological explanation:

'when the Jew is so broken and crushed that he has nothing to say, then he does not feel ... this is not silence [*harishah*] but rather muteness [*ilmut*] like the mute who has no power of speech.'[17] Muteness reflects spiritual weakening, a sign of psychological breakdown; but this radical withdrawal from engagement with the world is the personal responsibility of being an individual Jew. It is different from faith. He writes of 'broken and crushed' individuals, but Rabbi Shapira also believes God has the power to lift the community out of its muteness: God brings solace, refuge from the horror, even though refuge lies in faith, thoroughly expressed within the self. In September 1940, shortly after the murder of his family, he speaks of how the 'inner strengthening of will' turns 'evil into good,' in a Warsaw where evil appears to be the cause 'of such great troubles.'[18] Shapira uses words like 'downcast, broken, bent to the ground full of sadness,' but refuses to allow those moods to negate his faith, to take the observation 'my whole life is gloomy and dark'[19] as a cause to turn away from God. Unlike Nadia S., Shapira never wavers in his profound faith in God's presence.

To the very end Rabbi Shapira believes that only God 'can rebuild what has been destroyed'; and bring 'Redemption and Resurrection.'[20] His position has nothing in common with the partisans' strong belief in the group's political power to fight German violence. To preserve faith, Shapira argues, one must transfer fear of the world to fear of God, to His divine power, the 'revelation of His kingdom ... His sovereignty.' During such moments of faith, a kind of inspired introspection, the self 'feels elevated and joyful.' To be in awe of the majesty of God, to fear the power of God as the divine creator of the universe, as the holder of supreme justice, 'is indeed pure ... a supernal fear, which elevates the person.'[21] To live in fear of the world, however, is to lower oneself, to be without moorings, deaf to the word of this divine power. Shapira refers to the alienation of the self, to its being lost in the 'scriptural sense,' disconnected from faith and God; but 'God will search for us and find us. He will ... rescue our bodies and our souls with great mercy and beneficent acts.'[22]

God, then, has the power to heal what Shapira describes as a mass schizoid universe – disconnection, numbness, absence of empathy – trampling down any traditional selfhood and annihilating the will to live. He talks about the prevailing 'turmoil and confusion ... . No day, no night ... the whole world lies upon us, pressing down and crushing, to the breaking point.'²³ But despair is not finality; in being God's 'beloved,' the suffering Jews possess within themselves, as a community, as Israel, a divine significance impervious to German power and destructiveness. '[S]ince God's yearning for me is in a measure larger than my self, then I grow to become a greater human being; I now overshadow my essence.'²⁴ By refracting my presence in God's, I become more than I am; God creates me in His 'divine hands'; therefore, my corporeality is holy and sanctified by 'divine speech,' the Word or Torah. My very soul, in the hands of God, cannot be touched by the corporeal or secular world.

Shapira recognizes the broken spirits, despair, but insists that the Jew's responsibility lies in maintaining self-control while confronting *din* [judgment], the terrifying power of severity, and holding fast to faith. The more rigor the self exerts in overcoming this terror, the greater the likelihood of divine salvation. God may even receive one's suffering as a gift, and the self that offers up to God pain and affliction demonstrates an act of devotion, love, a calling, even if the gift of suffering were not freely chosen. Further, the body's wasting away can be understood as sacrificial suffering, a signifier of faith, 'the diminishing of [our] body's substance, energy and mental capacity' is to be experienced as sacrificial 'and a revelation of His light, holiness and salvation.'²⁵ As can be imagined, resistance survivors had no sympathy for this position.

Sacrifice during the Holocaust takes on terrifying properties. Rabbi Shimon Efrati responds to a petitioner seeking absolution for inadvertently smothering an infant to avoid detection. (In underground shelters or other hideouts, infants' cries often endangered the lives of those trying to escape roundups, including parents, friends, and relatives. Efforts to prevent them crying, for example, holding a hand tightly over an infant's mouth, on occasion resulted in the death of the infant.) Rabbi Efrati responds: 'The man who did this should not have a bad conscience, for he acted lawfully to save Jewish lives.'²⁶ At the time, however, such situations placed an enormous burden on sanity, on maintaining any sense of oneself as a human being, still alive in a human world; smothering one's child or the child of a friend or relative to save one's own life would raise terrifying moral dilemmas, in addition to the unimaginable grief of the parent. Where does the self place the guilt and sorrow? How to explain it? Such questions preoccupied Rabbi Shapira and Jewish law during this period and posed daily tests of faith. Situations where one might cause the death of another whether intentionally or not faced ghetto inhabitants every minute of every day: smuggling, participation in resistance, hiding from the Germans, stealing food, avoiding selections. Rabbi Shapira, in his elaboration of the conditions of faith, adapted theology to these murderous environments. Post-war absolution is one thing; but grief and guilt at the moment required a rabbinical response too.

Domination: patterns of injustice

After January 1942, the ghettos turned into vast collection centers for eventual transport to the death camps. In the Warsaw ghetto, by the winter of 1942, more than half a million had been herded into an area originally intended to house scarcely more than 20,000. It was no part of human experience or the human spirit to believe that an entire culture wishes to destroy you and your children solely on the basis of your biological existence. As early as spring and summer 1942, rumors were circulating about the death camps and extermination centers; but psychological and physical dislocation brought on by transport, the strangeness of unfamiliar circumstances, differences in culture and belief amongst the Jews themselves, made it impossible to sort out truth from rumor, much less organize a sustained resistance to terror.

The practices of domination in the ghetto (brought on by German policy) often turned Jew against Jew; the rich against the poor; the Jewish police against the indigent and unemployed; adults against children; smugglers against official administrators; the *Judenrate* against Zionists and communists; those with influence against those with no connections; crooks and thieves against families and individuals trying to survive.

But where was one to turn? Leaving the ghetto, except in a work brigade or with rare, official permission, constituted a capital crime. Poles, more often than not, turned over Jews found outside the ghetto to the German authorities. Jews in work brigades found themselves subjected to indignities and exploitation; laborers, organized into brigades and taken outside the ghetto, were treated like vermin. Mortality rates soared in the labor groups; random shooting, accidents and illness killed many in and around the work sites. One observer writes during the period of transports to Sobibor and Treblinka:

> 'The streets contained pitiful sights in these ghastly days of July, August and September 1942. Just before the *Aktion* ended, five tiny children, two- and three-year-olds, had been sitting on a camp bed in the open for 24 hours; presumably their mothers had already been taken to the *Umschlagplatz* [deportation site]. The children cried piteously, screaming for food – doomed.'[27]

The immediacy and prevalence of death placed extraordinary demands on theology. Shapira addresses individual cases of sacrifice; but his theology never embraces the human status of infants and children. Nor can it be expected to. Infants do not experience the 'powerful yearning to surrender life for the sake of the sanctity of His blessed name'; the five-year-old child screaming for his mother is incapable of raising 'up all his sense[s] for the greater glory of God.' Nor does the fifteen-year-old girl feel the ecstasy of the transcendence of the body, where 'sensory awareness disappears [while] feeling and corporeality are stripped away so that [consciousness] feels nothing but pleasure.' Nor does the eleven-year-old boy, yet to be *bar-mitzvah*, believe that, because of his suffering and the lice covering his head, and his body so weak he can hardly walk, faith will purge 'his sins and purify him, so that he might' attain a state of salvation.[28] It would not be fair, however, to condemn theology for refusing to take into account the inability of the young to understand faith or to reach the level of transcendence that Shapira demands.

To have faith against the backdrop of the Holocaust meant that the self could resist the German assault, at least spiritually. Theological or secular faith, it did not matter; as long as belief could resist the German definition of Jewish identity, one could act. Faith also protected consciousness against the deadly muteness and numbness of dissociation. Intensely strong belief structures sustained identity and a sense of selfhood, if not survival; and political faith, at least in the early days, energized armed resistance. Theological faith strengthened the self against degradation, forged an internality capable of withstanding physical abuse, and situated consciousness in a universe of meaning that was rapidly being destroyed by the Germans.

A young man ready to submit himself to the gas chambers to save his friend, a noted Torah scholar in his village, asks a rabbi, also in Auschwitz, for guidance *and* religious sanction. The rabbi wonders why he wants to do this and the boy responds that the world needs his friend's learning and erudition. He saw himself as ignorant and foolish with no potential to be a great scholar. He had witnessed the death of his parents and sisters; now, alone in Auschwitz, he had no desire to live. By substituting himself for his friend, even though it is a suicide, he will be performing a good act, a *mitzvah,* worthy of God's praise. Even though the boy pleaded, Rabbi Meisels refused to religiously sanction the substitution. He found nothing in his recollection of Talmudic law that would justify such a suicide.[29] But repeatedly rabbis faced moral contradictions that stretched very thin the interpretation of Talmudic law. Marek Edelman, a leader of the Warsaw ghetto uprising, relates:

'And when the baby was born, the doctor handed it to the nurse, and the nurse laid it on one pillow and smothered it with another one. The baby whimpered for a while and then grew silent. This woman [the nurse] was nineteen years old. The doctor didn't say a thing to her. Not a word. And this woman knew herself what she was supposed to do.'[30]

This happened while Germans were murdering elderly patients on the first floor of the hospital.

No matter what the cost in economic self-interest, the German policy was to kill all Jews. Daniel Goldhagen argues, 'the only groups of currently employed workers whom the Germans killed *en masse,* necessitating the closing down of manufacturing installations, were Jews. *Operation Erntefest* [Operation Harvest Festival], just one example of such a self-inflicted German economic wound, took the lives of 43,000 Jewish workers for whom they had no substitutes.'[31] The number of Jews in Poland working in German installations fell from 700,000 in 1940 to 500,000 in 1942. By June 1943, the number stood at around 100,000. Factories employing non-Jewish workers, slave and otherwise, never closed; non-Jewish workers, while exploited and abused, escaped systematic annihilation. Goldhagen draws the conclusion that the Germans' lack of economic rationality towards the Jews reinforced 'the already considerable evidence that they viewed and treated the Jews as beings apart, as beings – whatever else was to be done with and to them – ultimately fit only to suffer and die.'[32] Rabbinical authority responded to the German genocidal policy by shifting the ground and context of theological interpretation; what guided much in their thinking had to do with compassion in the face of assault and tragedy. Yet, the theology, particularly Rabbi Shapira's belief that faith residing in mind and soul was untouchable, possessed a sense of unreality. For example, Shapira's sermons argued that the soul constituted a space of freedom impervious to brutalization. The resistance, of course, took a very different position;[33] and given the rapidly disintegrating conditions inside the ghetto, it is understandable why resistance fighters paid very little attention to arguments about the interiority of faith.

The following leaflet, circulated by a clandestine labor group in the Lodz ghetto, demonstrates how the principles of genocide dominated work and living places six months before the bureaucratic plan for the Final Solution emerged at the Wannsee Conference in January 1942, how defining German biological hatred of the Jew was even in the early days of the occupation of Poland.

'July 20, 1941

All of us remember the terrible epidemic of last summer – dysentery, typhus, other diseases from which thousands died. We had no way of saving them. Then during the last three months, 3,000 Jews died of hunger and cold. Now we are threatened by a new epidemic. This one could be even more terrible than the last one, for people are weakened. They don't have the energy left to withstand its ravages. Our yards and streets are filled with refuse and garbage; the toilets overflow with excrement, and most of them are broken.'[34]

Genocide had one purpose: to kill. Primo Levi writes:

'Nothing obliged German industrialists to hire famished slaves.... No one forced the Topf Company (flourishing today in Wiesbaden) to build the enormous multiple crematoria in the

Lagers; that perhaps the SS did receive orders to kill the Jews, but enrollment in the SS was voluntary; that I myself found in Katowitz, after the liberation, innumerable packages of forms by which the heads of German families were authorized to draw clothes and shoes *for adults and for children* from the Auschwitz warehouses; did no one ask himself where so many children's shoes were coming from?'[35]

In the Maidanek and Auschwitz memorials, there are entire rooms filled with children's shoes.

But faith promised deliverance; holding out hope even for the body since deliverance might occur before death. Shapira: 'Who knows how long this will go on? Who knows if we'll be able to endure it... . The person is overwhelmed with terror, the body is weakened, one's resolve flags. Therefore, the most basic task is to strengthen one's faith, to banish probing questions and thoughts, trusting in God that He will be good to us, saving us and delivering us.'[36] But what does this 'choice,' this surrender to faith, the injunction against 'probing questions' of God, mean in the context of imminent peril, in the fetid workshops of Lodz, the tenement buildings housing starving and sick workers? Lawrence Langer writes: 'The Germans buried people twice, once before their death, and once after, and this is perhaps the most vicious of their many crimes. How is it possible to bury a man while he is still alive? How is it possible to make innocent Jews feel that they are murderers too? The Germans managed to find a way.'[37] The teenager in the Lodz ghetto who wrote the following lived not in faith but in despair:

'May 15, 1944

I have been saying lately that the inhuman state of mind we are in may be best proved by the sad fact that a Ghettoman, when deprived of half a loaf of bread, suffers more terribly than if his own parents had died. Was ever a human being reduced to such tragic callowness, to such a state of mere beastly craving for food? ... it is only German artistry in sadism which enables this, which makes it possible... . We are exasperated, despairing, dejected and losing hope. Our hunger grows stronger continually; our suffering is unimaginable, indescribable; to describe what we pass through is a task equal to that of drinking up the ocean or embracing the universe.'[38]

Disabling will through apathy, confusion, inaction – literal psychological collapse – began, Langer argues, long before physical death. Undressing in a gas chamber signified a final step in a process that had debilitated the will and annihilated the spirit. Parents watched their children die; children witnessed the death of their parents. By the time the victim reached the gas chamber, the Germans had transformed death into an integral part of life.

But even in the midst of this living death, which Shapira witnessed, the meaning of the Jewish commandments appears to the self as revealed truth; it is not necessary, Shapira argued, that such commandments be experienced or 'explained,' since no meditation or rational cognition stands between truth and justice. Tragedy, the unexpected, the uncanny, the thousands of dead bodies in the streets of Warsaw are 'without reason; but faith too is above reason, so that when we bind ourselves with a perfect faith to God [Who is] above reason, then even the *hukkah*-type calamities are transformed into sweetness.'[39] While murder of Jews is incomprehensible, beyond reason, faith guiding the self through suffering and calamity brings consciousness closer to the divine. For Shapira, this provided consolation. More problematic, however, and unknowable, is how much consolation faith provided to the Jewish families waiting for deportation to Treblinka and Sobibor.

Shapira's strenuous pleas that faith not be shaken; that it is the highest goal of consciousness, indicates that possibly just the opposite was happening in the ghetto. Diaries suggest religious belief declined not because of secular or theological debate; it was not an argument about God and Jewish identity between secularists and rabbis that provoked the decline in faith. The primary physical and psychological attack on religion came from oppression and starvation, causing an emotional collapse that left the self without affective content. Consciousness for many wandered in a numb no-man's land, outside the orbit of faith and reason. To retain faith in an environment of horror required a super-human leap of faith; an extraordinary act of will. In the face of tragedy the leap drained much of the human will. Shimon Huberband recites the words of a 22-year-old Jewish woman: the devastation in Warsaw 'was not, regrettably, an empty dream, nor a mad fantasy, or an evil tale, but naked and bitter reality.'[40]

Theological argument, however, placed the Word of God on a higher level than the world of action. Even though the Germans might assault Jews, if God through his signs and infinite wisdom revealed the word 'salvation,' salvation as a concrete historical fact may have arrived. But until that time, salvation remains on the purely spiritual plane. Presumably the 'Word' is revealed through the rabbis and their universe of holy utterance; but the arrival of salvation in the temporal world may be 'delayed.' To speak the Word of God, to pray, is a *mitzvah*, a form of action and of protest. To believe in eventual salvation and to continue believing in the face of devastation becomes a theological proposition *and* a psychological act providing solace and hope for the soul. Speaking prayers, holy observance, brings salvation closer, hence the emphasis on prayer, the repetitive uttering of holy words in Hasidic theology.

The critical *theological* question for the believer, Shapira maintained, should not be affected by the secular quest of how to deal politically with the real, immediate world, with the organization of workshops, work brigades, death camps, since performance of ritual, observance of prayer and dietary law constitute forms of action – the only route to realizing the Word of God, of Torah. Reason, understanding, explanation, the empirical reality of oppression, have no bearing on faith. What is required, then, of the Jewish community is a 'radical surrender to the divine will,' not to politics or the real. In Shapira's view it is a terrible mistake to think that reason can act as a protection against slaughter. Since all secular thought is sullied and twisted by experience, thought, if it is to be at all pure, unaffected by evil, needs to begin from a foundation in faith and God. It is not the individual who thinks alone, but faith that thinks through the self.

Soul-death and faith

The technology and bureaucracy of annihilation enforced a psychological fragmentation so vicious that ghetto inhabitants accepted the sight of dead and dying bodies in the gutters, streets, on sidewalks, as part of daily normality. Often children as young as five or six would haul the wagons carrying the nightly toll of dead bodies to mass graves, or excrement from makeshift privies to trenches dug in the ghetto. How is one to interpret Shapira's faith in the midst of all this? Spirit struggled in an infinite variety of sadistic contexts. Because of the omnipresence of death, parents, on occasion, deserted their children or gave them over to German guards or Jewish police making selections. In the Lodz ghetto during the infamous selection where the Germans demanded 10,000 children under ten, elderly over 60 and the sick as the price for sparing 20,000 adults, this twisted mélange of German efficiency, death and sadistic enthusiasm destroyed will. One father left his infant daughter for the Germans:

> 'When I came to the hospital ... I deserted her. I, her father, did not protect her. I deserted her because I feared for my own life – I killed ... I can't write – I deserve to be punished – I am the one who killed her. What punishment awaits me for killing my own daughter ...? I killed the child with my own hands.'

The Germans effectively killed this man *and* his faith without directly assaulting his body. By 1944, if he had not died from hunger, disease or cold, the gas chambers of Auschwitz or the gas vans of Chelmno would have destroyed him. Mooka's father psychically sustained 'life,' but it was living death, not sustained by faith:

> 'I walked off with Anya [his elder daughter] but I left Mooka behind. Instead of hiding with her in the cellar or in the toilet, I put her in a clothes basket, and she gave herself away with crying. Naked, barefoot, miserable – my dear child, it's me, your father, who betrayed you, it's me, driven by selfishness, who did nothing for your salvation, it's me who spilled your blood.'[41]

In Lodz, Oskar Rosenfeld describes a scene of resistance to the *Kinder* selection:

> '[A] child is torn away from a young woman by a *Feldgrau* [a German soldier so described by the field-gray of his uniform]. "Let me have my child or shoot me." *Feldgrau* pulls out [his] revolver. "I will ask you three times if I should shoot." He asks three times. The reply is always *yes*, and he shoots the woman down.'[42]

But examples like this are rare; ghetto inhabitants sink into catatonia, numbness. Shapira condemned this as a 'rebellion against God'; but that was not the case. Numbness or dissociation defined the ghetto's psychological reality. It was the consequence of the German brutalization – all too successful in inducing spiritual death. Even at a time when God's presence is hidden, Shapira argues, the self must continue to believe in Him: 'Everything that comes from Him ... is good ... just. Suffering embodies, in His hidden purposes, God's love for Israel.'[43] Shapira refuses to countenance as legitimate any questioning of God's intent: 'Faith is the foundation of everything.' No one can expect to know God's intent. 'How can we expect, with our minds, to understand what He, may He be blessed and exalted, knows and understands?'[44] After all, suffering is not unique to the Holocaust, since 'at the time of the destruction of the Temple, and at the fall of Beitar ... there were [sufferings] such as these.'[45] When one surrenders 'his soul,' when the self merges with God through faith, the consciousness of suffering is transcended; 'he will believe with perfect faith that everything is [transpiring] with justice and with the love of God for Israel.'[46] Oppression should turn consciousness towards the 'holiness' within, since holiness is more powerful than mind or reason. Oppression should glorify faith.

Even the Jew who strays from faith retains the possibility of returning to God's devotion, of submitting once again to the holy presence within. And in an observation clearly directed at the deteriorating spiritual conditions in the ghetto, Shapira writes, even if the self lies 'prostrate, like a stone, with mind and heart arrested' by savagery, even if 'improper thoughts assail' consciousness, it is essential to find one's way back to faith.[47]

The constant reference in sermons to the turning away from faith is powerful evidence suggesting how fragile the hold of religion had become in the community. Shapira time and again in his homilies of 1942 returns to the contrast between the theme of abnegation and purification. He speaks of having to 'nullify ourselves,' an extraordinary spiritual demand in Warsaw at that time. To consider oneself 'a separate being with his own mind' is to be 'outside' of God; but to be devoted to Him is to realize 'our minds are naught' and that if 'God made things happen this way, that's how it should be.'[48] German barbarity may push the self into a place where consciousness does not 'feel [its] faith'; but even though the Jew may be incapable of experiencing God's 'joyous state,'[49] faith as a fundamental relation to God never disappears. It is there, like one's 'stomach, heart or lungs'; the self may not be aware of the existence of faith, just as we take for granted the operation of the lungs or heart.[50] We carry within us as a given of the soul 'the allotment of love and faith ... as our forefathers' legacy.' And just as we cannot add more lungs to our body, we cannot possibly add more love and faith.

Faith for Rabbi Shapira, then, possesses an involuntary presence; it is there, whether we are in touch with it or not. It is outside of us but it is also inside, although it may be deeply buried and 'outside of our field of awareness.' When the self reaches for it, when consciousness during 'a state of slackness and weakness' transcends itself, even though a person may not 'perceive' faith, he is nevertheless a 'believer.'[51] Shapira acknowledges that disbelief may arise, it is inevitable; but questioning God should not be taken as a sign that faith has disappeared altogether. There may be momentary lapses, but return to faith always exists even in the face of strong evidence opposing faith. What will assure the community's salvation is not action in the world, not politics, but the binding power of faith and God's covenant with His people.

7
The Silence of Faith Facing the Emptied-out Self

One can admire Rabbi Shapira, but in the force of the German assault, psyches were crushed; and psychological collapse and with it the disintegration of spirit often, but not always, preceded physical death. Terror might erode the boundary between inside and outside, the world of the soul and that of the body. Yet, as some diaries and Jewish law written during this time describe it, in this struggle to survive, the individual could say to himself, 'Over my soul they have no dominion.' But equally powerful was the impression of souls dying.

A survivor described to me ghetto life in Warsaw:

> 'You have no idea what it was like, the filth, hunger, dead bodies all over the streets, thousands of children, many covered head to foot with lice, begging or wandering aimlessly. I remember one woman, walking down a street, stumbling over bodies, murmuring something like, "Mendel, Mendel," her arms stretched out in front of her, her eyes crazed; she had no shoes, her clothes hung off her body in tatters.'

In the camps and ghettos, 'the only thing that you can think of is that you're hungry.'[1] A survivor in Krakow told me that his father had been fortunate enough to be placed on Oscar Schindler's list, thereby assuring protection against the Germans. But this man, a respected doctor in the Krakow Jewish community, left behind in the Plazow labor camp his wife, son and two daughters. It would have been insensitive of me to ask what he thought of his father for doing this; but evidence of his bitterness appeared in the fact that he changed his name to a Polish name, married a Polish Catholic woman and refused to bring his children up as Jews, thoroughly rejecting his father's name and religion. Yet, he visits Auschwitz every year and goes out of his way to take foreign visitors to Auschwitz where he, as a survivor, has the dubious privilege of being able to drive his car onto Auschwitz-Birkenau grounds. One can only imagine what this ten-year-old boy felt when he watched his father leave him in the labor camp, to find safety in Schindler's factory.

Surviving was brutal for the person lucky enough to survive, but the process of survival could mean abandoning entire families. Miraculously, Dr. B.'s entire family survived the war. Dr. B.'s parents emigrated to Israel; Dr. B. stayed in Poland. I asked him if he had visited his parents: 'Once or twice I went to Israel.' One wonders how the terrified child would respond to Rabbi Shapira's concept of faith and devotion, and the eternal presence of God's care and concern.

Lawrence Langer recounts Anna G.'s description of an event on the ramp at Auschwitz. A ten-year-old girl refused to go to the 'left' (which meant death) after the initial selections. The child, seized by three guards who held her down, screamed to her mother to help her, to stop the guards from killing her. One of the guards approached the mother and asked if she wanted to go with her daughter; the mother said no. The eye-witness describing this scene said to the interviewer: 'Who am I to blame her? What would be my decision in a case like this?'[2] Sidney L. witnessed the death of his parents and several siblings. Against this kind of assault, to maintain an inner world of faith impervious to the outer world of body or event would be psychologically impossible. The power of the assault and the attack on his will and faith appear in Sidney L.'s description of how he survived: 'In all these things that happened – I played a very small part in everything that happened. There were very few things that I initiated, or planned out on this. This is how it happened; it took me from here and put me there ... . It was not my plan, it was not my doing.'[3]

For many survivors, chance takes the place of God. A 17-year-old boy, with working papers, believes that the Germans will let his brother accompany him to a labor camp. But instead, the SS insist his brother go to the 'left.' 'I know it's not my fault, but my conscience is bothering me. I have nightmares, and I think all the time that the young man, maybe he wouldn't go with me, maybe he would survive. It's a terrible thing; it's almost forty years, and it's still bothering me.'[4] Or Sally H.: 'I'm thinking of it now ... how I split myself. That it wasn't *me* there. It just wasn't me. I was somebody else.'[5]

The impact of such a psychological assault transformed the self, emptied it out. For example, Bessie K., a young wife in the Kovno ghetto in 1942, tried to smuggle her baby into a work camp; the Germans seized the infant. 'And this was the last time I had the bundle with me.' Part of 'her,' her self, identity and being, died. 'I wasn't even alive; I wasn't even alive. I don't know if it was by my own doing, or it was done, or how, but I wasn't there. But yet I survived.' To survive, she says, she had to kill feeling inside herself; part of her had to die. In the boxcar to Auschwitz: 'I was *born* on that train and I *died* on that train ... but in order to survive, I think I had to die first.'[6] In Irene H.'s words, 'The truth is harsh and impossible to really accept, and yet you have to go on and function.' What life showed was 'a complete lack of faith in human beings.'[7] Another survivor even rejects luck as the agent for her survival: 'I had determined already to survive – and you know what? It wasn't luck, it was stupidity.'[8]

Survivors rarely speak of the intense feelings generated by the terror; although in my interviews with resistance survivors there appeared to be a willingness, even a need, to recapture those feelings in their narratives. But in Langer's transcripts of the Yale Fortunoff Library's collection of survivor testimony, it is almost the exact opposite, as if all feelings including faith had been suspended by the daily demands of survival:

> 'It is difficult to ... talk about feelings ... we were reduced to such an animal level that actually now that I remember those things, I feel more horrible than I felt at the time. We were in such a state that all that mattered is to remain alive. Even about your own brother or the closest, one did not think.'[9]

Here Vernon describes his exhaustion: 'You ask me if we talked about faith; we were too tired at the end of the day even to talk.'

Yet, a simple act like lighting a candle on Yom Kippur became a significant even venerable event in the camps or in the forests. The only reality that possessed day-to-day emotive content was remaining alive. Alex H. remarks that 'fraternal caring is a major measure of civilized conduct,'[10] but caring as a communal act could be sporadic and haphazard during the Holocaust. 'Sure, I cared for other people's emotions, but not very often,' Vernon observed, 'the governing law was "every man for himself."'

Responses that in any kind of normal environment would be considered natural, in the Holocaust environment possess lethal potential. In a factory, an SS officer came up behind Luna K. and cocked his revolver. Her mother was sitting opposite her. Both women remained completely silent, no words, no protest (although the mother's face turns chalk-white), even though each was convinced he would pull the trigger. Luna K. heard a click; in fact, it was the gun trigger striking an empty chamber. The officer had literally run out of ammunition. 'So nobody says anything ... it wasn't worth taking my life, so he just walked out. So now you can understand why people were quiet. If my mother said a word, I wouldn't be here today.'[11] Yet, in the recollections of resistance fighters, it is not muteness that defines forest life, but constant action and noise. The more noise, the safer they were. Sonia Bielski: 'If we could speak loudly amongst each other, we were safe.'

Martin L. describes a state of mind pervasive throughout the ghettos and camps, an immediacy or actuality that literally flattens consciousness. 'When you see a lot of deaths, your mind gets numb, you can do nothing ... . Your humanity is gone. You're speechless.'[12] The deadly potency of muteness recurs throughout the diaries and recollections, the sense of will-lessness, the ineffectiveness of speech and volition. Contrast this to Rabbi Shapira's faith in the power of the word; speech as the key to divine revelation. But Shapira understands the situation all too well and allows that there may be moments of silence or muteness between God and the self; and between God and the community. That does not, however, mean that faith or God has disappeared. But apparently this was a matter of some debate and concern in Warsaw, since Shapira returned to this theme throughout his sermons.

For many survivors, the Germans took away will and civilization. Langer's analysis is grim: civilization, will, energy fall apart; 'gradually, gradually, you become a different person. And you do things that you would *never* think you'd do – and you do it.'[13] What Langer calls the 'disintegration of basic life' extends to the psyche; George S. describes one woman whose child was discovered and taken by the Germans; her despair appears as a compulsion to reveal to the German authorities the hiding places of other children. The very human premises of what Rabbi Shapira called goodness disappear against the force of reality, which, distorted by the presence of evil, defines all value and power. Indeed, there is evidence to suggest that many Warsaw Jews believed evil had destroyed God.

Many survivors, of course, stayed with their families and tried to save them; but eventually reality collapsed hope and expectation. Leon H. witnessed the death of each member of his family; he believed that by staying near them, he might serve as a protective shield. But each died; finally, he tried to defend his brother, the last surviving member of his family, but his brother died in his arms in a camp. He feels morally responsible for events he could not control: 'We envied the dead ones.'[14] A mother hid the decaying corpse of her five-year-old child under the bed for several weeks in order to keep claiming the child's food rations. Moses S. describes a concentration camp as a place where they take you 'to die and die and die.'[15] Langer sees the dissociative psychological process as a mode of survival: the 'paradoxical killing of the self by the self in order to keep the self alive.'[16] Yet, to kill the 'self' means to kill the affective or emotional self, what Winnicott calls the core self: the self of feeling and identity.[17] It is to adopt a 'false self' system (a phrase borrowed from Winnicott); but the surviving false self is more like a mask disguising a dead inner reality, so thoroughly terrorized, that to allow feeling inside would jeopardize the very survival of consciousness and being. One can admire Shapira's desperate attempts, through words, to fight this deadly feeling of dissociation and detachment. But Langer's survivors stress time and again how they were driven into muteness.

Survivor testimony consistently returns to this theme: killing the self not only to defend against an intolerable reality but to assure the possibility of physical survival. Also selves engage in actions that never would have been imagined prior to the Holocaust: stealing, handing over children, indifference to death, discarding of traditional moral values. The consequence, in Langer's view, psychological indifference, testifies not to the endurance of faith or the redemptive power of God, but, in the words of one survivor, 'These people come back, and you realize, they're all broken, they're all broken. Broken. Broken.'[18] For these survivors, Rabbi Shapira's homilies take on the status of dysfunctional fantasies. But it is a complicated story; survivors in their new lives retained belief in their Jewishness. They joined synagogues, celebrated *Bar* and *Bat Mitzvah*, lit candles on Friday night. Yet, these actions seemed to be separate from the knowledge of who or what had saved them. It was not God, but in the case of resistance survivors, guns and action. It would be wrong, then, to argue that faith had been killed by the Germans; it would also be wrong to see Rabbi Shapira's faith celebrated in what survivors describe as the continuing assault on spirit. Faith had not been killed by the Germans; it had been challenged and dealt with brutally. But for the resistance survivors – and their testimony is of course quite different from those found in Langer's book – the sense of membership in an historical community, the emotional and religious core of a Jewish identity, and pride in that membership, appeared in religious and theological observance after liberation, and during the Holocaust, in the resistance communities themselves. Women expressed a more profound belief in God than the men; and much of that belief had to do with the association of God's will with natality, the biological link between generations.

Survivor testimony, however – and this includes much in the partisans' narratives – resonates with the memory of selves being broken in the ghetto, of faith in God being nowhere present in day-to-day efforts to evade capture, in the disintegration of moral limits, and the unrelenting self-absorption of individuals desperately attempting to stay alive. A survivor of the Lodz ghetto: 'When you're hungry, it gets to a point where you don't mind stealing from your own sister, from your own father ... . I would get up in the middle of the night ... and slice a piece of bread off my sister's ration. Now I – you would never picture me, and I can't even imagine myself doing that now. But it happened.'[19] Another victim of the Lodz ghetto told me during an interview in Warsaw that he saw families fighting each other to pick up a scrap of bread from the street. Given this breakdown of ethics and morals, Rabbi Shapira's sermons appear as a profound invocation of God to contain or transmute the natural or biological conditions of survival, to overlook this very real Hobbesean decline into incivility. A survivor of the Plazow labor camp told me in an interview in Warsaw that after one month in the camp, the word 'God' never entered his mind.

For Langer, the victims embody a 'monument to ruin,' a striking contrast to Holocaust theology attempting to find meaning or redemption in the destruction. For these survivors, there can indeed be no God after Auschwitz. Langer looks at survivors not as a testimonial to God or a harbinger of the state of Israel, but as victims broken beyond imagination by the German project of mass slaughter. He also, unjustly I believe, criticizes the term 'spiritual resistance,' a concept he finds absent in the survivor testimonies. But for many survivors, particularly resistance survivors, survival itself testifies to a spiritual endurance protecting the self from madness and absence of will.

All survivors I interviewed emphasized the connection between their identity as a Jew and the rescue of the self from madness. It is in this context that I find the concept of spiritual resistance to be most persuasive, although I should add that my approach to what spiritual resistance means has been influenced by the partisans' narratives. Langer, however, maintains that survivors 'demur virtually unanimously [about spiritual resistance] when it is raised by an interviewer.'[20] He argues that the concept of spiritual resistance does not 'require any control over one's *physical* destiny.'[21] Yet, in my interviews survivors repeatedly expressed how connected their spiritual wellbeing – by which I understood them to mean maintaining their sanity – was with the violence of their partisan units, although that sense of spirit had nothing to do, they insisted, with faith in God or in the belief that God had directed their actions. Langer interprets spiritual resistance far too narrowly. To postulate resistance on the level of spirit may be seen in the sheer act of survival itself, spirit maintaining focus amidst the very real possibility of madness and *will* not being crushed by the physical harshness of the surroundings. It required will, discipline and spirit to survive, yes, but it also required luck. However, the stories themselves testify that luck and food were not the only factors. Langer underestimates the power of even remembering in the midst of such suffering the *Bar Mitzvah* of one's child, standing under the *huppah* with one's bride; Sabbath meals, the fantasy of revenge. Surely in this sense, spirit never succumbed completely. Langer: 'perhaps that explains why we retreat to spiritual resistance – to reestablish a veneer of respectability for situations in which harsh necessity deprives the individual of the familiar dignity of moral control.'[22]

But 'harsh necessity' and 'moral control' may have nothing to do with spiritual resistance: both phenomena defined the conditions of the camps, although resistance fighters certainly possessed more 'moral control' than camp inmates. The capacity of the self to retain its memory, its sense of identity, its power to distinguish between good and evil; its wish for the practice of ritual, even if it is reciting a silent prayer – are these not evidence of spiritual resistance and not a 'veneer of respectability'?

Langer quotes Emmanuel Ringelblum, who refers to the 'complete spiritual breakdown and disintegration caused by unheard of terror ... the enemy does to us whatever he pleases.'[23] Yet, Langer refuses to accept Ringelblum's assessment of such Jewish action as 'quiet passive heroism'[24] and sees nothing heroic about passivity. In his view oral testimonies conclusively demonstrate how desperate individuals, brutalized by Nazi terror, lost all perspective on the maintenance of moral boundaries and spiritual dignity. He quotes Sol R., recalling a friend killed during an air raid over a concentration camp. His friend 'was always loaded with bread, and here he was lying dead, and I grabbed his bread and I gorged myself ... I've been choking on that bread ever since.'[25] Yet, does this mean that Sol R. had suffered spiritual collapse? Or are the destruction of moral limits and the annihilation of spirit two discrete phenomena? Langer refuses the possibility that psychological survival, the endurance of the soul's connection to its Jewish identity, even in the face of the breakdown of 'moral limits,' signifies acts of spiritual resistance. Are not Rabbi Shapira's faith and courage, the commitment of hundreds of rabbis and their congregations to protect sacred Torah scrolls, significant as acts of spiritual resistance? Langer is not making moral judgments about behavior. Quite the contrary: he wants to elicit from the testimony a view of surviving without the moral perspective of retrospective embellishment. In Eva K.'s words: 'My fate pushed me, you know. I [could] not help myself.'[26] Or Chaim E.: 'On the other side, you didn't have any choices. You just were driven to do whatever you did... . You do whatever you have to do, from *other* people.'[27] Yet Langer fails to appreciate the psychological moments of spiritual resistance or even to consider that uttering *Kiddush haShem* might suggest transcendence over the facts of barbarism, even though prayer had no efficacy in preventing murder.

For Chaim E., the Germans drained agency and morality from action: '[W]e were not individuals, we were not human beings, we were just robots where we happened to eat and we happened to do things. And they kept us as long as we have any function ... . Now if the function was not good, we don't need you, [we] destroy you.'[28] In a universe where 'today was already better than tomorrow,'[29] moral rules, pre-Holocaust traditions, suffered drastic revision. Chaim E. had been interned in Sobibor and participated in the escape attempt in which 75 prisoners succeeded and survived the war. Many hundreds more, however, were killed during the attempt and in the aftermath. Chaim E.'s motives, Langer points out, were practical not spiritual, and certainly not spiritual in Rabbi Shapira's sense. For Chaim E., 'only the survival for your skin, that's what counted.'[30] But why discount the possibility that at the moment of liberation and even during the period of planning, there may have been a sense amongst the escapees that even though death might be imminent, their action constituted an example; that Jews locked in death camps need not die without action and resistance. What their action meant, what it signified as an act of resistance and affirmation, must have been a consideration – not only in the midst of those planning the Sobibor escape, but with small bands of fighters in large and small ghettos. Is this not 'spirit'? Did not these groups think collectively and at times with a common spirit which moved them away from despair and apathy?

In Langer's analysis of testimony, it was not spiritual resistance that kept survivors alive but chance, luck, moral transgression and a supreme grasp of the practicality of life. But the very act of survival meant that spirit itself had survived; that it had resisted the German efforts to crush it. Langer's judgment, then, might be too harsh. Although Rabbi Shapira's homilies might not be appropriate to understanding the state of mind of those who survived, they possess significance in suggesting that a vision of the world was present that provided comfort and meaning to thousands for whom death was an absolute certainty. Langer discredits such a vision as lacking any effect; but his judgment may in this instance be too quick, too abrupt. Spiritual resistance was ineffective against the bullet, gas chamber, death pit. But it did provide a space for spiritual identity, something the Germans could not touch.

Rabbi Shapira's theology and, it would be fair to say, the unrecorded utterances of rabbis who perished, moved on two levels: to give hope to those not yet killed and to provide spiritual help for those waiting for transport or dying in ghetto streets – consolation in the midst of the horror. Those who died possess no voice; but if Rabbi Shapira's teachings and his presence in the lives of those thousands who knew him, or knew of him, provided an affirmation, a justification of faith, then the very reality of disintegration and moral collapse that survivors describe may have had an opposing presence; maybe – and who can ever know – at least a few of those who suffered a miserable death in the gas chambers, mass burial pits or isolated forests, received the bullet, gas or fire with the knowledge that the enemy could not take away from them their belief in God, that while God had failed to protect them from death, they knew that redemption lay on the other side.

If theology gave one person a meaning, an explanation, the hope of vengeance and redemption, then while the body may have gone to its death abjectly, the spirit may have remained untouched. In this sense, one can, unlike Langer, speak of and admire spiritual resistance; it is not a vapid phrase, an idealization of memory, a retreat to the fantasy of heroic action. It is action understood through the utterance of words; it is a form of resistance that appears over time in the Talmud. Prayer affirms the belief that while Amalek, the incarnation of evil, may take away the body, he cannot take away the soul. Few witnesses record what this belief meant at the time of death; but what is clear is that the theological vision did not fail. True, it never formed the basis for any mass political resistance, and given survivor and diary accounts of psychological and spiritual collapse, any impact the theology might have had in saving lives was minimal. But it would be wrong to impugn the concept of spiritual resistance only on the basis of survivor accounts; while the breakdown between inner and outer may explain the mass apathy and silence, the few accounts we do have of *Kiddush haShem* – sanctification of the name of God – at the moment of annihilation would suggest a faith not dead or dysfunctional, but a set of beliefs held by many and undoubtedly taken to their deaths by many. That, I would argue, constitutes spiritual resistance – the only kind of resistance available to millions of religious and devout Jews for whom partisan warfare possessed no realistic possibility.

Political alternatives existed, but one had to be able to understand and conceptualize their possibility, to think politically, to organize escape and resistance. The spiritual universe of East European Jews was incapable of making that kind of leap. Orthodox theological strategy in dealing with oppressors historically had involved accommodation, bribery, waiting it out, aligning with political factions sympathetic to the Jews. In Poland, where over half the Jewish population *practiced* Hasidic Judaism, theological explanations possessed meaning, context *and* authority – acting as a refuge or a psychological 'safe' zone in the emotionally and physically battered self. Thus, a rabbi tells his congregation: 'We will march straight to that place where rest the righteous for whose sake God has permitted the world to endure.'[31] Or, 'the sanctity of the *Shoah* martyrs pierced the heavens, and the Almighty redirected the course of Jewish history. The process of Redemption began to unfold.'[32] This is spiritual resistance. Emmanuel Ringelblum in a diary entry dated February 27, 1941, notes that the rabbis of Krakow had been sent to Auschwitz because they had attempted to intervene with the Germans to stop a mass deportation. Should not this be considered resistance of both the body and the spirit? For the vast majority of East European Jews, secular political ideologies had been shunned.

The following declaration typifies the power of faith as a position in this world.

> 'Listen to me, brothers and sisters … . We are the children of the people of God. We must not rebel against the ways of the Lord. These, our sufferings are meant to precede the coming of the Messiah. If it was decreed that we should be the victims of the Messianic throes, that we should go up in flames to herald the redemption, then we should consider ourselves fortunate to have this privilege. Our ashes will serve to cleanse the people of Israel who will remain, and our death will hasten the day when the Messiah will appear. Therefore, brothers and sisters, let not your spirit falter. As you walk into the gas chambers, do not weep but rejoice.'[33]

It would not be unreasonable to assume that such imagery might have had a positive effect in the general human environment of sunken, dull eyes, the disintegration of communication, withdrawal into the self, apathy in the face of assault and deadness in emotional reactivity. It is not unusual, for example, in clinical literature to find descriptions of abused children whose psychological universes resemble those of the inmates of the ghettos and camps. If we see such examples as carrying Torah scrolls, singing religious songs, rabbinical utterance, as *action* responding to despair (uttering sacred words in the Talmud is considered a form of action), then these forms of spiritual protest, rather than being ineffective, appear as a last, desperate effort to lift a psychically disoriented community out of radical emotional withdrawal and catatonic displacement.

Kiddush haShem was not a response that posed real, political consequences, like, for example, Mahatma Gandhi's passive resistance, sitting down in front of armed troops; but not even underground fighters or partisans could transform the political environment of genocide and the violence of the Final Solution. To have done that would have required a Jewish political organization, across Europe, far stronger and with far more political authority and access to weapons than existed among the splinter groups of Jewish theological and secular factionalism. The spiritual reliance on sanctified words as a response to brutalization evolved as an active effort to utter what amounts to a silent 'no,' a passive resistance taking place entirely within an oppressed community and invisible to the aggressor. And in analyzing the Jewish response to the Germans, it is essential to keep in mind the relentless assault against men, women, children and infants, who possessed no political language or resistance.

Rabbi Chaim Ozer urges students fleeing from the Nazis to 'dedicate yourself to studying diligently ... in this time of *churban* [disaster], the sounds of Torah study have been dimmed. It is our obligation, we who still have the ability to pursue our studies, to dedicate ourselves to it with added energy and diligence.'[34] Here he addresses spirit, not politics – not the political reality of annihilation, but a spiritual universe, which prayer affirms, saving souls from extinction. Hasidic rabbis believed God would rescue the Jews; but in the absence of God's rescue efforts, *Kiddush haShem* replaced the coming of the Messiah. Faith struggling against reality, verbal injunctions like 'don't open your mouth to Satan,'[35] prayer recitations, or repetitive moments like 'springing' on one's toes[36] before attempting to jump from a transport train, were thought to have power. Faith sanctified God's goodness, no matter what the external events revealed. Or, according to Rabbi Ozer: 'When a battle is waged against Jews, it is also being waged against *haShem* and against Torah – can there be any doubt, then, as to which side will persevere in the end?'[37]

There is no evidence to suggest that younger Jews became more religious during the Holocaust, although some took considerable risks in attempting to observe traditional holidays such as Yom Kippur or Passover. Cultural identity could be reinforced and affirmed, although *ritual* practices like bathing in the *mikvah*, lighting candles, baking *matzah* at Passover were not the same as *theological* practice, for example, the study of Torah. That dimension of faith, study as action, seemed to intensify in some ghettos such as Kovno.

Martyrdom for God has a long tradition in Jewish history; it is central to the teachings of the Talmud. 'You may kill my body, but not my soul.' Some survivors, however, felt differently. Abraham K.: 'I hated the orthodox rabbis. They sat in their studies and rendered judgments; we came to them and asked for their help, money, or influence with the *Judenrat* in helping the resistance. They just said nothing in the Torah could justify that course of action.'

It is doubtful, however, if financial support for buying weapons would have drawn more religious Jews into the fold of active partisan or underground movements. While younger Jews tended to be more mobile, and secular Jews less inclined to religious dress, habits and practice, the Germans made no distinction about who was to be annihilated. No belief, concept or class status assured a greater probability of survival. Surviving in the ghettos depended mostly on contingency, will and the ability to withstand disease and starvation and avoid the omnipresent selections and random killing that made the difference between life and death. But for those who perished, the theology of *Kiddush haShem* may have made a difference. History, tradition, culture and theology made it impossible for the Jewish community to make the leap from *Kiddush haShem* to political *resistance*. So sacrifice for God and the historical community took on extraordinary significance. Rabbinical authority followed the ancient words of Maimonides: 'When Israel is forced to abolish their religion or one of the precepts, then it is the duty of the Jew to suffer death and not violate even any of the other commandments, whether the coercion takes place in the presence of ten Jews or in the presence of non-Jews.'[38] To die for Israel, therefore, is purposive, meaningful. 'If one is enjoined to die and not to commit the transgression and suffers death and did not transgress, behold, he has sanctified the Name of God.'[39] Partisans, underground fighters, of course, saw it differently. But outside the gas chambers, parents made children recite *Viddui,* confession of sin, to be in proper spiritual accordance with the sanctification of the Name of God.

It would then be wrong to see *Kiddush haShem* as symptomatic of cultural abjection or psychological death. People believed God would eventually take revenge; to be able to utter the name of the Lord on the threshold of the gas chambers, to project a consciousness of faith in the face of the enemy, suggests an aspect of the Jewish self that, given the presence of German power and its effects, should be accorded respect as a form of resistance. Ritual practice continued, on rare occasions, even in the death camps; for example, *tfilin* became prized articles of possession in Auschwitz, usually worth three or four portions of food in trade. A survivor remembers a *Succah* erected in a corner of a camp workshop; others recall *matzah* prepared during Passover, a *shophar* blown during *Rosh haShanah.*

But side by side with *Kiddush haShem* is the story of emptied-out selves, the diaries describing the catatonic, the aimless wanderings in the street, the lost and homeless, those whose being-in-the-world progressively comes to resemble madness, forms of human behavior that in a more 'normal' environment would be labeled psychotic or dissociated. That too needs to be acknowledged. Facing total psychological collapse, some rabbis counseled active resistance; for example, Rabbi Yitzhak Nissenbaum of Warsaw exhorted Jews to survive and resist the oppressors. 'Jews should do everything – by flight or bribery – to live.'[40] A somewhat different picture was given to me by a survivor of the Warsaw ghetto, David L.: 'Yes, one or two rabbis, after the fall of 1942, counseled resistance; but the vast majority did not. Most rabbis were dead by summer of 1942.'

8
Law and Spirit in Terrible Times

With a long tradition in Judaism, responsa (rabbinical judgments on issues pertaining to moral and community life) and their commentary constitute much of Talmudic content. During the Holocaust, rabbis wrote judgments regarding marriages, living arrangements, the burial of the dead and attendance at burials; they recited prayers at funerals, presided over ceremonial occasions, officiated in the *shul*, and supervised the slaughtering of animals for meat. The responsa constitute an intimate glimpse of how rabbis organized moral life and the assumptions used to guide individual and group behavior, particularly regarding compliance with German regulations. *Judenrat* leaders, as well as ordinary citizens, consulted the rabbis, who provided moral coherence for a universe rapidly disintegrating under the force of German rule. Many of the surviving responsa deal with petitions for remarrying without proof of death of the former spouse; during and after the war, they provide detailed glimpses into the human and social devastation imposed on Jewish life. The following example is typical of responsa regarding remarriage.

David wished to marry a woman he had known for a few months; his wife and five children had disappeared on a transport to the East; her husband was shot in a mass grave in a forest near the ghetto. David lacked proof of death, but knows his wife's destination was Auschwitz. Should she be considered dead? Evidence of death, of course, is nonexistent; but many people in the ghetto saw her pushed into the boxcar. David asks for the rabbis to rule for a presumption of death.

The Kovno *Judenrat* brings before Rabbi Abraham Shapiro, Chief Rabbi of Kovno, a request to decide if the council should hand out yellow labor cards; cardholders will be exempt from any future selection. Those not holding cards face an almost certain death. The *Judenrat* office is besieged by hundreds of people demanding labor cards. The *Judenrat* asks the Chief Rabbi to decide whether labor cards should be distributed. 'If the decree to destroy an entire Jewish community has been determined by the Enemy,' Shapiro writes, 'and through some measure or other it is possible to save part of the community, its leaders are obliged to summon up their spiritual strength and take upon themselves the responsibility of doing whatever needs to be done to save a part of the community.'[1] The philosophy of 'saving the remnant' guided Shapiro's deliberations, even though, in retrospect, it could be argued that these kinds of decisions fell into Primo Levi's 'gray zone.'

In Auschwitz, a young boy asks a rabbi to bless his decision to substitute himself for another boy in a group selected for the gas chambers. A group of several hundred children between the ages of nine and fourteen will be killed the following morning. He wants rabbinical sanctification for the act, but is prepared to go ahead even if the rabbi refuses judgment.

Every day rabbis faced the prospect of bending traditional sources to the needs of the moment and to write decisions consistent with the imperative of survival. In this respect, the responsa indicate how close the rabbis were to the moral and psychological hammers inexorably destroying the Jewish community. For example:

> 'Rabbi, my only son is in that cellblock [in Auschwitz]. I have enough money to ransom him. But I know for certain that if he is released, the *kapos* will take another in his place to be killed. So Rabbi, I ask of you a *she'elah le'halakhah u'lema'aseh* (a question which demands an immediate response to an actual situation). Render a judgment in accordance with the Torah. May I save his life at the expense of another? Whatever your ruling, I will obey it.'[2]

Rabbi Meisels refuses to make a judgment, arguing that the situation is so unprecedented that, without proper source books, it would be impossible to make a 'reasoned' decision. But the man insists: 'Rabbi, you must give me a definite answer while there is still time to save my son's life.' The rabbi again declares nothing in the Talmud sheds light on this situation; the father takes this to mean that the rabbi will not sanction a ransom attempt. 'Rabbi, this means that you can find no *heter* [permission] for me to ransom my only son. So be it. I accept this judgment in love.' But the rabbi argues his silence should not be construed as disapproval: 'Beloved Jew, I did not say that you could not ransom your child. I cannot rule either yes or no. Do what you wish as though you had never asked me.'³ And the father's final response:

> 'Rabbi, I have done what the Torah has obligated me to do. I have asked a *she'elah* [question] of a *rav* ... . In your own mind; you are not certain that the Halakhah permits it. For if you were certain that it is permitted, you would unquestionably have told me so. So for me your evasion is tantamount to a *pesak din* – a clear decision – that I am forbidden to do so by the Halakhah. So my only son will lose his life according to the Torah and the Halakhah. I accept God's decree with love and with joy. I will do nothing to ransom him at the cost of another innocent life, for so the Torah has commanded.'⁴

God is never questioned in the *Halakhah*; His 'judgment' remains pure. 'For the believer there are no questions; and for the unbeliever there are no answers.'⁵ Yet, suicide and death rates dramatically rose. In January 1942, 5,000 died on the streets in Warsaw or in tiny apartments crowded with fifteen or twenty families. But the faith of the rabbis in the ancient biblical injunction never wavered: 'The secret things belong unto the Lord our God, but the things that are revealed belong unto us and to our children forever, that we may do all the words of this law' (Deuteronomy 29: 28). In Lodz, distraught parents, losing children to starvation, illness or random selections, throw themselves out of three-story windows; men and women, disheveled, covered with lice, roamed the streets asking to be shot. In Warsaw, totally withdrawn, disturbed children, in the gutters or door wells, lay in their own waste; the elderly dropped dead on sidewalks. These conditions had no effect on rabbinical judgments. The rabbis who were still alive affirmed what God means as majesty and power. The following from the *Hekhalot Rabbati*, an ancient mystical text, describes this unshakeable faith: 'Wonderful loftiness, strange power, loftiness of grandeur, power of majesty.'[6] While the *Hekhalot* is a mystical text from a different rabbinical tradition than the *Halakhah* or traditional law, both insist on an unquestioning, uncritical attitude towards God.

In 1939, outside Lublin, a group of Hasidic Jews forcibly assembled by the Germans was ordered to sing a light, popular Hasidic melody. Someone, however, initiated the more solemn song, *Lomir zich iberbeten, Avinu shebashomayim* [Let us be reconciled, our Father in heaven]. An observer at the scene describes the reaction:

'The song [initially ordered by the Germans], however, did not arouse much enthusiasm among the frightened [Jewish] masses. Immediately Glovoznik [the troop's commander] ordered his hooligans to attack the Jews because they refused to comply fully with his wishes. When the angry outburst against the Jews continued, an anonymous voice broke through the turmoil with a powerful and piercing cry: "We will outlive them, O Father in Heaven." Instantly the song took hold among the entire people, until it catapulted [them] into a stormy and feverish dance. The assembled were literally swept up by the entrancing melody full of *devekut* [strong spiritual adherence] which had now been infused with new content of faith and trust.'[7]

A rabbi exhorts the few thousand left in the Warsaw ghetto after the mass deportations of 1942 to guard 'against dejection and depression, and to support ourselves in God,' even when God demonstrates no interest in protecting His people. 'True,' he says, 'this is very, very difficult, since the suffering is too much to bear.' By now, almost every survivor in the ghetto had lost a family member; death wagons daily picked up piles of corpses; the few thousand left lived in fear of selection and transport; food supplies were almost nonexistent; fuel had disappeared; apartments were unheated and had no running water or functioning toilets; human feces lay in apartment hallways. The stench of rotting and diseased flesh permeated the ghetto. The rabbi continues: 'However, at a time when many Jews are burned alive sanctifying God, and are murdered and butchered only because they are Jews, then the least we can do is to confront the test and with *mesirat nefesh* [dedication] control ourselves and support ourselves in God.'[8]

One rabbi in Warsaw in the spring of 1943, however, staked out a different theological position. Rabbi Menachem Zemba rejected the notion that death itself sanctifies God's name: 'I insist that there is absolutely no purpose nor any value of *kiddush haShem* inherent in the death of a Jew. *Kiddush haShem* in our present situation is embodied in the will of a Jew to live.' To live in Warsaw in the spring of 1943 meant, according to Zemba, not just to survive, but to fight. 'This struggle for aspiration and longing for life is a *mitzvah* [religious imperative] [to be realized by means of] *nekamah* [vengeance], *mesirat nefesh* [extreme dedication] and the sanctification of the mind and will.'[9] Zemba provided both a theological and political contrast to Rabbis Shapira, Oshry and Meisels.

Compare Zemba's position with the proclamation of Rabbi Yehezkiah Fisch who, it is reported, prior to entering the gas chamber of Auschwitz, cried out with a joyous clapping of hands: 'Tomorrow we shall meet with our Father,' or Rabbi Shem Klingberg, in Plazow, who prayed before his death, 'May it be thy Will that I have the privilege of atoning for all Jews.'[10] *Kiddush haShem* had little to do with Rabbi Zemba's vengeance, *nekamah*. Hasidic theology saw the Holocaust as 'the descent for the sake of the ascent,' the darkness of suffering that precedes redemption.[11] In the Jewish messianic tradition, central to Hasidic belief, each generation creates its own martyred Messiah to prepare the way for the final redemption. Periods of suffering, therefore, in the history of the Jewish people, beginning with the binding of Isaac, that have long preceded the coming of the final Messiah, comprise integral parts of the Hasidic story; and it is probably true that many Hasidic Jews went to their deaths believing the Holocaust to be part of that messianic hope, *Hevle Mashiah*, another agonizing trial preparing the Jewish people for the coming of the final Messiah.[12]

Nothing in this mystical theology would suggest a natural affinity for more activist forms of political resistance, although themes in Hasidic texts suggest that Zionism and Hasidic theology have affinities in common. 'God,' the theology proclaims, 'will bring you together again from all peoples where the Lord your God has scattered you ... . And the Lord your God will bring you to the land.'[13] Rabbi Levi Yitzhak: 'God took the Israelites out of Egypt in order for

160 *Jewish Resistance during the Holocaust*

you to inherit the land. With your coming to the land, you will have achieved completeness.'[14] But unlike the Zionists' political position, the Hasidim believed prayer – not action – would make that union into a reality.

Mysticism and faith: action as belief

Hasidic tradition believes that, by fulfilling God's will, the self joins with holiness. In striving for *devekut*, by merging with God, the community of believers sanctifies holiness. The self seeks the divine through prayer, which is active joyousness, a celebration to God. While all human souls have the capacity to serve God and be with Him, much Hasidic ritual is devoted to dissolving the self, as a way of reaching the divine state of *Ayin* or total submersion in God. In Hasidic religious practice, ritual, particularly singing and dancing, moves the self closer to the realization that God's justice is absolute and unquestioned. For example, the Piasezner Rebbe proclaims: 'Although nothing shall remain of me but bones, they will still continue to proclaim "Lord, who is like unto Thee!"'[15] Moral law originates in heaven and emanates from God's will; evil and suffering possess agency in God's plan; suffering becomes purposive and reflective of God's intent. Acceptance of God's will, with bountiful love, no matter how obscure that will seems to be, serves as the prime requirement of faith. This relationship to God is not rational but a complete giving over of the self to God's will. *Emunah* [belief] defines the intimate connection between man and God, an unflinching compact, absolute in its meaning and intent.

The historical and popular confusion over why Jews so 'willingly' went to their death overlooks this aspect of Hasidic theology practiced by millions of Central and East European Jews. The Belzer rabbi whose firstborn son was killed in a synagogue burned down by the Germans, expresses his loss in terms of religious belief: 'It is indeed a kindness of the Almighty that I also offered a personal sacrifice.'[16] The Hasidic self is literally ruled by an absolute commitment to God's will and a horror at even the possibility of disobeying God's injunctions regarding unquestioned faith: 'Suffering glorifies the self, sanctifies the community and affirms the chosen people's place in history ... the greater the darkness and suffering on the eve of Sabbath, the more brilliant the light of the Sabbath.'[17] It is a sign of religious sacrifice. '[I]f we truly conceive the ultimate purpose of all as is, we would most certainly accept all of suffering with longing and love, since it is by means of these beatings that the Name of our Creator will be magnified and sanctified.'[18] Rabbi Yehezkiah Fisch, the Matislaker Rebbe in Hungary, asks on the eve of the deportation of Hungarian Jews to Auschwitz: 'Is it not worth suffering prior to the coming of the Messiah?'[19]

Ministering to the flock occurred in many different ways in the Jewish spiritual sector; the following responsa, for example, indicates the desperate pressure the Jews faced. It is also curious why Hannah Arendt's harsh condemnation of Jewish traditionalism in *Eichmann in Jerusalem* failed to acknowledge the effort of spiritual leaders to respond to human suffering.

In 1942, Rabbi Oshry was asked by a member of his congregation a question relating to forced labor. Labor brigades were proscribed from bringing any food into the camps; further, the Germans strip-searched all prisoners returning to the camps to make sure no food had been smuggled in. Because his children were starving, one man, on one of his work assignments, smuggled six pieces of bread in his pants. The Germans discovered the bread, then beat and mutilated his genitals. When the wounds healed, he consulted the rabbi and assured him he would still be attending morning and evening services, even though he was in a great deal of pain. His question to the rabbi involved the mutilation of his genitals; he believed this would preclude his being honored as a *kohen*, and therefore receiving ritual privileges. A passage in the Bible (Deuteronomy 23: 20) prohibits mutilated persons from participating in any official capacity in services. The rabbi assured him participation still was possible and yes, he could be included in the services as a *kohen*. Later, the petitioner was shot by the Germans.

This man's wish to act with his congregation, to be spiritually alive in the face of his terrible physical trauma, shows a self actively moving away from passivity or acquiescence, even as he suffers the death of his children and the mutilation of his genitals. Such examples must be considered when making interpretations about spiritual resistance. This wish to be *kohen* is not 'active' in a strictly political sense, as Rabbi Zemba's call for vengeance in the Warsaw of

1943 was. But the responsa shows a self struggling against spiritual death. Or in Berkovits' words:

'There were really two Jobs at Auschwitz: the one who belatedly accepted the advice of Job's wife and turned his back on God, and the other, who kept his faith to the end, who affirmed it at the very doors of the gas chambers, who was able to walk to his death defiantly singing his *Ani Mamin – I believe.*'[20]

Yet, doubts remain about the truth of Berkovits' assertion, 'If God was not present for many, He was not lost to many more.'[21] I put this proposition to a survivor of Auschwitz. Her response: 'I never stopped being a Jew, but I hated God every minute of my stay in that place.' When a starving inmate of Mauthausen, near death, eats human flesh to survive, God's presence may be difficult to find. How does a child of seven or eight find 'truth' or 'fulfillment' in *Kiddush haShem*? Martyrdom is the furthest thing from the mind of the young girl who cries out outside the gas chamber, 'Please God; let me live'; who bitterly laments never having experienced love or motherhood; the father who watches his four-year-old hauled away by the Germans thinks not about God's sanctification but about his own powerlessness. For the mother who jumps from an upper floor after watching the last of her children die from starvation, God lacks immanence and refuge. If a Jew, Berkovits argues, 'is able to accept his radical abandonment by God as a gift from God that enables him to love his God with all his soul, "even when He takes his soul from you," he has achieved the highest form of *Kiddush haShem.*'[22] For many, however, that was a very difficult proposition to accept.

Even Ringelblum's admiration for orthodox Jews degraded by German soldiers becomes for Berkovits a sign of the ennobling effects of *Kiddush haShem*. But the theology paints only part of the picture; the diaries draw another, as with the father in Lodz who, forced to turn over his two-and-a-half-year-old daughter to the Germans, can no longer bear to think of himself as a human being.[23] A survivor, Miriam K., lost her entire family: 'For me, God has no existence; He is neither alive nor dead; he just was not there.' Ringelblum rarely speaks of religion in his diaries; for him, too, God remains hidden. Yet the surviving responsa apparently indicate terrible moral conflicts facing the Jewish community and the few remaining spiritual leaders attempting to adapt Jewish law to the circumstances. Law and prayer were all the rabbis had to offer.

Spiritual refuge or psychological disintegration: who can ever know?

A thirteen-year-old boy is whipped mercilessly by a guard in Auschwitz. Suffering scores of lashes, the boy utters nothing; no sound, no cries, just silence. The theology sees such resilience to pain as deriving from the child's faith in Torah. For the rabbi, 'they can take my body but not my soul. They have no authority over the soul.' But it might not have been faith that allowed the boy to be whipped and left bleeding; it may have been a dead self produced by a deadly reality. Another explanation might argue that the numbness and dissociation literally annihilate the child's emotional universe, making him impervious to pain. A survivor writes:

> 'I knew that I had to act soon because once malnutrition set in, I would lose the will to fight. Next would come the indifference to my surroundings, vacant eyes, swelling at the ankles, and the slow descent into oblivion. I had seen the pattern thousands of times in the ghetto and was seeing it again in people around me.'[24]

Massive assault and abuse do tremendous damage to the psyche; Joseph Horn: 'I was in shambles. I felt less worthy than an animal. A stray dog could walk the streets unmolested, but the whole might of the Third Reich would be enlisted to hunt down me or any other Jew who didn't cooperate in our own destruction. I began to realize that we are all destined to die.'[25]

This is not to diminish the effects of rabbinical compassion or adaptive law; but to idealize suffering as a testament to God, as well as to criticize Jews for not resisting more, misses the profound psychological reality of death that consumed the Jews long before they ever reached Auschwitz. For the orthodox, the issue was not questioning the action itself or even attempting to understand genocide and mass death. Whatever questions were addressed to God were directed to effects and His tactics. It violated faith to question God's intent. Rabbis quoted the Bible: 'It is a time of agony unto Jacob, but out of it he shall be saved' (Jeremiah 30:7). Orthodox Jews never asked, 'But at what cost?' In the Maidanek memorial, in a barracks to the side of the camp, lie over 800,000 pairs of shoes taken from Jews; in a corner, piled high, one finds tens of thousands of children's shoes. The mound of bones and ashes containing the remains of almost a million Jews standing at the top of Maidanek in a gigantic stone urn may represent not a martyrdom for Israel but a meaningless sinking into nothingness, not to be remembered as a sacred victory of the Lord, but a concrete reminder of what happened to six million. 'A man must pronounce a blessing over evil just as he pronounces a blessing over good' (Berakhot 9: 5); but the ashes and bones of Maidanek represent for many survivors a victory of evil – pure and simple. Was it 'the rock, His work is perfect; all his ways are justice' (Deuteronomy 32: 4)? Or was it, in the words of a survivor: 'I cursed God when I watched my little girl die; I have yet to set foot in a synagogue again.' Joseph Horn recounts the following conversation with a friend in the ghetto: '"This is a terrible time to be a parent," Mrs. Psherover said. "When you can't protect your child from such depravities, perhaps it is best to put an end to it."'[26] When Jews cried *zaddik-ve-re-lo* (why should the innocent suffer), did God send consolation to the soul or did the self's inner being remain empty, uncomforted? I asked Abraham X, a survivor of Mauthausen:

> 'Since liberation, life for me has been a waking death; all my family were killed. What I have done since then is to just walk through life. God means nothing to me. I can't understand those who speak of having faith after Auschwitz; there is no God; where is God for my murdered sons and daughters; for my wife, my parents?'

And those who speak about the righteousness of suffering, or how it is ordained for Jews to suffer?

> 'That means nothing to me; for years, after the war, I walked around thinking about how to kill and murder as many Germans as I could find. That hatred kept me going. There was no place for God inside my soul. Who was it, I don't remember now, the survivor of the Warsaw uprising, who said, "If you lick my heart, it

would poison you?" That's how I felt and probably still do. I remember the line in *Psalms* [91: 15]; "I will be with you in distress." I don't believe in God. While we were in distress, God was somewhere else, maybe helping the Germans kill my children.'

Not all survivors believe in the sanctity of suffering and many would disagree with the ideas expressed in the following theological text:

'They said to his wife: "It has been decreed on your husband that he is to be burned, and upon you to be killed." She recited this scripture, "A faithful God, never false."

They said to his daughter: "It has been decreed upon your father to be burned and upon your mother to be killed and upon you to perform labor." She recited this scripture: "Wondrous in purpose and mighty indeed, whose eyes observe" [*Jeremiah* 32:19].'[27]

I read these selections to Abraham X; he was furious. 'Tell that to me while I say *Kaddish* for my dead children, or while I think of the possibilities they might have had, the grandchildren that never were. Everything I am in my old age, everything that I was, lies inside my head, a photograph album as real as the very days when I had them, when all of us were alive.' 'God's justice – His deeds are perfect ... a faithful God ... there is no perversity' (Deuteronomy 32: 4) is not Abraham X's justice.[28]

Abraham X's bitterness moves against the most powerful strains of *Midrashim* thought: 'Precious are sufferings [for] just as the covenant is established by virtue of the land, so too is the covenant established by virtue of suffering.'[29] Precious, however, might not be the right word to describe those psychologically and physically decimated by German power. Joseph Horn: 'Now, I could hardly bear to look at him [a former friend]. His cheekbones were sunken, and the suit he wore was dirty and crumpled. His eyes were bloodshot. He was incoherent as to the whereabouts of his family – I think he still believed the canard about tilling the land in the Ukraine.'[30] Death quickly follows the loss of faith. 'When [inmates in the camps] gave up hope, their eyes became vacant. Some would go on the electrified wire for a quick death. Most would lie down on the bunk for the night and expire without a whimper' – a common occurrence, not only in the camps but in the ghettos.[31]

In Jewish theology, to study the commandments, the law, or its interpretations becomes, in a religious sense, equivalent to carrying out the practice of devotion, righteousness. The Jew by *studying* acts for God. It is a religious dialectic; self speaking with Other/God, where doing lies literally in the speaking (prayer and meditation). And while, for example, Aaron Bell admires this now, at the time the only action that mattered was the violence of retribution. For him and the other resistors I interviewed, the violence of their resistance communities constituted the only trustworthy form of prayer. But for those like Rabbis Shapira and Oshry study brought the self closer to God, to contact and intimacy with the Divine Will, a mystical joining, a coming together and coming apart, in the Law. Praxis is the actual study of the written and oral law. To act means to be 'with' God, to render respect to God and to be and become worthy as a Jew, a student of Torah. It is a state of 'being' and a way of life that places political ideology and political action not only in a secondary position but in a place that possesses no emotive or evaluative significance. Political action in this theology has no role in improving or demonstrating the worthiness of the self or one's ethical place in the world. Worthiness and, therefore, value depend on how close study takes one to God. Perhaps this explains some of the political and ideological quietism of orthodox and Hasidic Jewry.

In the Torah divine justice is higher than compassion and certainly lies considerably higher than political action. Compassion signifies preference for one being over the other; it requires that God make a choice, discriminating between sufferings, granting compassion to some, but not to others. But in the theology, God makes no such discriminations amongst people. Divine justice affirms *only* God's majesty; all are equal in God's eyes; no one can claim preference, not even those who suffer, because compassion discriminates between 'pains' and judges some pain to be worthier than others. What matters to God are His Laws – nothing else; and His justice lies in the inviolability of His Laws. To be in a relation of faith to God is not to demand compassion but to obey, without question, His Laws, a *fear* of Heaven, to accept what is a manifestation of God's justice. To demand compassion of God means that one asks God to explain; but the revelatory power of God lies not in explanation or rationalization but in the immanence of His Laws. If the believer knows the Law and accepts it, God's justice is also immanent. God need not show compassion or explain its absence; nothing is required of God, since God's authority and eternal righteousness guide the children of Israel.

What the partisan survivors consistently returned to was the notion that the Holocaust demanded a set of actions taking even more seriously than God himself the suffering of the human body and especially the suffering of children. But to have asked for such sustained violent resistance from the Jewish community, they also argue, misses the point. It is not that more could be done. All that could be done was done. In the face of the German onslaught and the ruthless efficiency of starvation and extermination, to have retained any kind of faith was itself a victory and transcendence. To have sustained any kind of violent resistance took extraordinary courage and luck, and a willingness to construct an ethics that would facilitate survival itself.

Yet, essential to the existence of both violent and spiritual resistance was the capacity of the resistors to withstand the radical inversion of meaning and value imposed by the German occupation and German racist policies. Resistance occurred against a backdrop of madness. Group madness and individual madness are quite different, but there is one very frightening similarity: the power of perception to literally transform reality to conform with the projections of the subject. For the Jews, German brutality was mad, incomprehensible; but for the Germans, extermination possessed a logic, a significance, in protecting the nation's public and biological health.

The resistor represents and affirms an historical sanity, a value embedded in history and the history of ethics, and in the case of the Jews, the word of the Torah; the resistor, whether the action be violent or spiritual, refuses the premises of mass murder and the murder of an identified and marked group simply because this group exists. That assumption – the Other must die because the Other exists – has no standing in an ethical history that grants dignity to human life. When the resistor denied either violently or spiritually the Germans' view of reality, that refusal embraced an affirmation and sanity that German rage could not touch or destroy.

Jewish corpses were referred to as *Figuren* by the Germans. A whole vocabulary was constructed round the killing: disinfection, special treatment, national therapy, and so on. To see the Other as refuse for sanitation, burning, cleansing, and to transform this into national action and policy is as insane as the individual schizophrenic conducting a delusional war in the privacy of a fantastic imagination. It is infinitely more dangerous because the fantastic *ideology* provokes the group into externalizing the madness, into making madness appear to be 'normal' and 'rational.'

Indeed, resistors, whether Jewish or Gentile, sought not only to save lives, but to preserve the vocabularies of the sane, the *human* text and spirit which German action destroyed. The lack of protest in Germany and the rest of Europe to the genocide of the Jews testifies *not* to indifference or lack of participation; plenty of evidence suggests otherwise. What little resistance to mass murder by non-Jews says is that the ground of sanity and insanity had radically shifted; that what a decade or so earlier would have been regarded as mad became, in the environment of the Third Reich, sane, rational sanitation policy with the objective of protecting the physiological and psychological boundaries of the Third Reich. When madness comes to be taken for the norm, when evil is seen to be rational state policy, not only has history been violated, but the very premises of human thought itself and the use of reason in the service of generative, creative and productive ends, have been blown away. What Jewish resistance, both violent and spiritual, accomplished was to remind the world and posterity that sanity and courage had not been completely annihilated.

Notes

Introduction

1. Rabbi Kalonymus Kalman Shapira, *A Student's Obligation: Advice from the Rebbe of the Warsaw Ghetto*. Trans. by Michael Odenheimer. Northvale, NJ: Jason Aronson, Inc., 1991.
2. For an interesting analysis of the kind of planning that produced the Warsaw ghetto and the death camps, see Gotz Aly and Susanne Heim, *Architects of Annihilation: Auschwitz and the Logic of Destruction*. Trans. by A.G. Blunden. Princeton, NJ: Princeton University Press, 2002.
3. Eliezer Berkovits (in *Faith after the Holocaust*. New York: KTAV Publishing House, 1973) raises important issues about the impact of the Holocaust on both the internalization and understanding of faith.
4. For a fascinating analysis of evolving German patterns of moral recognition, see Bronwyn Rebekah McFarland-Icke's perceptive and detailed examination of psychiatric nurses' participation in euthanasia and their adaptive moral positions towards killing the mentally ill. *Nurses in Nazi Germany: Moral Choice in History*. Princeton, NJ: Princeton University Press, 1999.
5. Irving J. Rosenbaum, *Holocaust and Halakhah*. Jerusalem: Ktav Publishing House, 1976, p. 116. For stories and tales filled with religious and spiritual references, see Yaffa Eliach, *Hasidic Tales of the Holocaust*, New York: Oxford University Press, 1982.

1 The Moral Justification for Killing

1. For a comprehensive review of Jewish resistance during the Holocaust, see Yehuda Bauer, *The Jewish Emergence from Powerlessness*. Toronto: University of Toronto Press, 1979. For a detailed account of the largest and most successful partisan unit, in the sense that over 90 percent of the members survived the Holocaust, see Peter Duffy's history *The Bielski Brothers: The True Story of Three Men Who Defied the Nazis, Saved 1200 Jews, and Built a Village in the Forest*. New York: HarperCollins, 2003. See also Reuben Ainsztein, *Jewish Resistance in Nazi-Occupied Eastern Europe; with a Historical Survey of the Jew as Fighter and Soldier in the Diaspora*. London: Elek, 1974; Yehuda Bauer, *They Chose Life: Jewish Resistance in the Holocaust*. New York: American Jewish Committee, Institute of Human Relations, 1973; Ronald J. Berger, *Constructing a Collective Memory of the Holocaust: A Life History of Two Brothers' Survival*. Niwot, CO: University Press of Colorado, 1995. Michael Elkins, *Forged in Fury*. New York: Ballantine Books, 1971; Charles Gelman, *Do Not Go Gentle: A Memoir of*

Jewish Resistance in Poland, 1941–1945. Hamden, CT: Archon Books, 1989; Meir Grubsztein, ed., *Jewish Resistance during the Holocaust*. Proceedings of the Conference on Manifestations of Jewish Resistance, Jerusalem, April 7–11, 1968. Trans. from the Hebrew by Varda Esther Bar-on et al. Jerusalem: Yad vaShem, 1971; Israel Gutman, *Resistance: the Warsaw Ghetto Uprising*. Boston: Houghton Mifflin, 1994; Isaac Kowalski, ed., *Anthology on Armed Jewish Resistance, 1939–1945*. Introduction by Yitzhak Arad. Brooklyn, NY: Jewish Combatants Publishers House, 1984; Vera Laska, *Nazism, Resistance and Holocaust in World War II: A Bibliography*. Metuchen, NJ: Scarecrow Press, 1985; Michael R. Marrus, ed., *Jewish Resistance to the Holocaust: Selected Articles*. Westport, CT: Meckler, 1989; Yuri Suhl, ed. and trans., *They Fought Back: The Story of Jewish Resistance in Nazi Europe*. New York: Schocken Books, 1975; *Flames in the Ashes*. A film by Haim Gouri, Jacquot Ehrlich; produced by Monia Avrahami. Teaneck, NJ: Ergo Media, Inc., 1987; Harold Werner, *Fighting Back: A Memoir of Jewish Resistance in World War II*. Ed. by Mark Werner. New York: Columbia University Press, 1992; Yad vaShem, *Studies on the European Jewish Catastrophe and Resistance*. Jerusalem: Yad vaShem, 1963–74. See also Simha Rotem, *Memoirs of a Warsaw Ghetto Fighter: The Past within Me*. Trans. from the Hebrew and ed. by Barbara Harshav. New Haven, CT: Yale University Press, 1994; Isaac Schwarzbart, *The Story of the Warsaw Ghetto Uprising: Its Meaning and Message*. New York: World Jewish Congress, Organization Department, 1953; *The Lonely Struggle; Marek Edelman, Last Hero of the 1943 Warsaw Ghetto Uprising*. Video produced and directed by Willy Lindwer. Audio-Visual Arts and Productions. Teaneck, NJ: Ergo Media, 1995; *Not Like Sheep to the Slaughter: The Story of the Bialystok Ghetto*. Video. Manor Prod. Ltd.; producer, Yigal Ephrati; written and directed by Adah Ushpiz. Teaneck, NJ: Ergo Media, Inc., 1991.

2. Hannah Arendt, *Eichmann in Jerusalem: A Report on the Banality of Evil*. New York: Viking Compass, 1974; Raul Hilberg, *The Destruction of the European Jews*, 2 vols. New York: Holmes and Meier, 1985.

3. See Ruth Bondy's sensitive and moving analysis of Jakob Edelstein, head of the *Judenrat* in Theresienstadt. *'Elder of the Jews': Jakob Edelstein of Theresienstadt*. Trans. by Evelyn Abel. New York: Grove Press, 1989; also, Isaiah Trunk, *Jewish Responses to Nazi Persecution: Collection on Individual Behavior in Extremis*. New York: Stein and Day Publishers, 1979.

4. For a comprehensive analysis of public health issues in Warsaw under the German occupation, see Charles G. Roland, *Courage under Siege: Starvation, Disease and Death in the Warsaw Ghetto*. New York: Oxford University Press, 1992.

5. See Etty Hillesum, *Letters from Westerbork*. Trans. by Arnold J. Pomerans. London: Cape, 1987.

6. For a particularly wide-ranging and revealing collection of contemporaneous documents and diary entries regarding the tragedy of ghettoization, see Alan Adelson and Robert Lapides, eds., *Lodz Ghetto: Inside a Community under Siege*. New York: Penguin, 1991.

7. For a comprehensive account of the uprising, see Yitzhak Zuckerman, *A Surplus of Memory: Chronicle of the Warsaw Ghetto Uprising.* Trans. and ed. by Barbara Harshav. Berkeley, CA: University of California Press, 1993.
8. See Sheva Glas-Wiener, *Children of the Ghetto*. Trans. by Sheva Glas-Wiener and Shirley Young. Fitzroy, Australia: Globe Press, 1983; George Eisen, *Children and Play in the Holocaust: Games among the Shadows*, Amherst: University of Massachusetts Press, 1988.
9. Cf. Chaika Grossman, *The Underground Army: Fighters of the Bialystok Ghetto*. New York: Holocaust Library, 1987.

2 Collective Trauma: The Disintegration of Ethics

1. Gilles Lambert, *Operation Hatzalah: How Young Zionists Rescued Thousands of Hungarian Jews in the Nazi Occupation*. Trans. by Robert Bullen and Rosette Letellier. New York: the Bobbs-Merrill Company, Inc., 1974. See also Lucien Lazare, *Rescue as Resistance: How Jewish Organizations Fought the Holocaust in France*. Trans. by Jeffrey M. Green. New York: Columbia University Press, 1996; Rafi Benshalom, *We Struggled for Life: Zionist Youth Movements in Budapest, 1944*. New York: Gefen Publishing House, 2001; Ellen Levine, *Darkness over Denmark: The Danish Resistance and the Rescue of the Jews*. New York: Holiday House, 2000; David Cesarani, ed., *Genocide and Rescue: the Holocaust in Hungary 1944*. New York: Berg, 1997.
2. Shalom Cholawski, *Soldiers from the Ghetto*. San Diego: A.S. Barnes & Co., Inc., 1980, p. 58. For an interesting account of resistance hardship, see Rich Cohen, *The Avengers*. New York: A.A. Knopf, 2000. Similar criticisms were leveled against Avraham Tory, Chairman of the Kovno Judenrat. For his own account of his experience, see Avraham Tory, *Surviving the Holocaust: The Kovno Ghetto Diary*. Ed. by Martin Gilbert. Trans. by Jerzy Michalowicz. Cambridge, MA: Harvard University Press, 1990.
3. Jack and Rochelle Sutin, *Jack and Rochelle: A Holocaust Story of Love and Resistance*. St. Paul, MN: Graywolf Press, 1995, p. 67; Yechiel Granatstein, *The War of a Jewish Partisan: A Youth Imperiled by his Russian Comrades and Nazi Conquerors*. Trans. by Charles Wengrov. Brooklyn, NY: Mesorah, 1986.
4. Adina Blady Szwajger, *I Remember Nothing More: The Warsaw Children's Hospital and the Jewish Resistance*. Trans. by Tasja Darowska and Danusia Stok. New York: Pantheon Books, 1990, pp. 35, 36. See also Chaim Aron Kaplan, *Scroll of Agony: The Warsaw Diary of Chaim A. Kaplan*. Trans. and ed. by Abraham I. Katsh. Bloomington: Indiana University Press, 1999.
5. Szwajger, op. cit., p. 39.
6. Ibid., p. 42; for a terrifying narrative about the fate of Jewish children, see Tadeusz Pankiewicz, *The Cracow Ghetto Pharmacy*, New York: Holocaust Library, 1987.

7. Ibid., p. 43. In Warsaw the death toll of children rose dramatically. Between January and August 1941, the figures were: 450, 800, 1,200, 2,000, 2,500, 4,000 and 5,600. Figures are from a diary written in the Warsaw ghetto, quoted in Rafael F. Scharf, ed., *In the Warsaw Ghetto, Summer 1941*, New York: Aperture Foundation, 1941, p. 103.
8. Ibid., p. 45. See also Laurel Holliday, *Children in the Holocaust and World War II: Their Secret Diaries*. New York: Pocket Books, 1995.
9. According to Szwajger (*I Remember Nothing More*, pp. 137–8): 'Children are always being born. Even in hiding places and in cellars. But they often die, and it is not always possible to save them. You have only your own hands, but you also need medicines and mother's milk, which in this case did not flow at all. And they mustn't cry. The landlords were afraid ... I had to carry the body of the newborn baby from the house in a cardboard box. A different mother's child, an older baby, a few months old, I carried out in my arms in its swaddling clothes. But I don't wish to remember how I had to drink vodka with the undertakers before they would bury the body somewhere under the wall ... There were live children too. Few of them were with their parents. They were already grown up, with that maturity of five- or six-year-olds which taught them that you must never cry, that you must never talk, and that almost all day you have to lie in bed, on a pallet. Lying on a bed all day in some dark hole, children soon stopped walking. It is a recognized illness, called *rachitis tarda*, or late rickets. But you have to see it to know what it is like when a twelve-year-old girl lies without moving, and even when she is allowed to get up, she is unable to stand on her legs. And she cries without voice.' See also Azriel Louis Eisenberg, ed., *Lost Generation: Children in the Holocaust*, New York: Pilgrim Press, 1982.
10. Sutin and Sutin, *Jack and Rochelle*, p. 66; cf. Henry Orenstein, *I Shall Live: Surviving Against All Odds, 1939–1945*. Oxford: Oxford University Press, 1987.
11. Faye Schulman, *A Partisan's Memoir: Woman of the Holocaust*. Toronto: Second Story Press, 1995, p. 112. For a fascinating account of Jewish women in the Holocaust, see Vera Laska, ed., *Women in the Resistance and in the Holocaust*. Foreword by Simon Wiesenthal. Westport, CT: Greenwood Press, 1983; and Judith Tydor Baumel, *Double Jeopardy: Gender and the Holocaust*. Portland, OR: Vallentine Mitchell, 1998; Nechama Tec, *Resilience and Courage: Women, Men and the Holocaust*. New Haven, CT: Yale University Press, 2003.
12. Zygmunt Kowalski, *Diary from the Years of Occupation, 1939–1944*. Ed. by Andrew Klukowski and Helen Klukowski May. Trans. by George Klukowski. Chicago: University of Chicago Press, 1953, pp. 222, 226. On July 17, 1941, an observer of a meeting of the Warsaw *Judenrat's* 'trans-settlement commission' notes, 'As leaders of our community you have displayed total lack of sensitivity, of human feeling toward the suffering of the poor, toward the agony of a starving child. You have treated even the dead, your victims, no better than a dog's carcass.' Cited in Joseph Kermish, ed., *To Live with Honor and Die with Honor!*

Selected Documents from the Warsaw Ghetto Underground Archives, O.S. (Oneg Shabbath), Jerusalem: Yad vaShem, 1986, p. 302.
13. William M. Mishell, *Kaddish for Kovno: Life and Death in a Lithuanian Ghetto, 1941–1945*. Chicago: Chicago University Press, 1988, p. 97; see also Malvina Graf, *The Krakow Ghetto and the Plaszow Camp Remembered*. Tallahassee: Florida State University Press, 1989.
14. Quoted in Meir Grubsztein and Moishe M. Kahn, eds., *Jewish Resistance during the Holocaust*. Proceedings of the Conference on Manifestations of Jewish Resistance. Jerusalem, April 7–11, 1968, Yad Vashem, 1971, p. 61. For a comprehensive review of Jewish youth movements, see Asher Cohen and Yehoyakim Cochavi, eds., *Zionist Youth Movements during the Shoah*. Trans. by Ted Gorelich. New York: Peter Lang, 1995.
15. Personal interview.
16. Quoted in Y. Gottfarstein, '*Kiddush Hashem* in the Holocaust Period', in Grubzstein and Kahn, *Jewish Resistance during the Holocaust*, p. 468.
17. Chaika Grossman, *The Underground Army: Fighters of the Bialystok Ghetto*. New York: Holocaust Library, 1987, p. 190. See also Albert Nirenstein, *A Tower from the Enemy: Contributions to a History of Jewish Resistance in Poland*. Trans. from the Polish, Yiddish and Hebrew by David Neiman; from the Italian by Mervyn Savill. New York: Orion Press, 1959; '*Not Like Sheep to the Slaughter*': *The Story of the Bialystok Ghetto*, video written and directed by Adah Ushpiz, Manor Prod. Ltd., 1991.
18. Grossman, op. cit., p. 203.
19. Ibid., p. 210.
20. Ibid., p. 282.
21. Ibid., p. 282; see also Puah Rakovsky, *My Life as a Radical Jewish Woman: Memoirs of a Zionist Feminist in Poland*. Trans. by Barbara Harshav and Paula E. Nyman. Bloomington: Indiana University Press, 2002.
22. Ibid., p. 399; see also Arnold Zable, *Jewels and Ashes*. New York: Harcourt, Brace and Company, 1991.
23. Ephraim Oshry, *The Annihilation of Lithuanian Jewry*. Trans. by Y. Leiman. New York: The Judaica Press, Inc., 1995, pp. 64–5. See also Lester Samuel Eckman and Chaim Lazar, *The Jewish Resistance: The History of the Jewish Partisans in Lithuania and White Russia during the Nazi Occupation, 1940–1945*. New York: Shengold Publishers, 1977.
24. Oshry, op. cit., p. 69.
25. Ibid., p. 74.
26. Ibid., p. 75.
27. Personal interview.
28. Klukowski, *Diary from Ten Years of Occupation*, p. 196.
29. Ibid. See also Jan Tomasz Gross, *Neighbors: The Destruction of the Jewish Community in Jedwabne, Poland*. Princeton, NJ: Princeton University Press, 2001. For a rare account of resistance inside Germany, where protest succeeded in rescuing Jews from a Gestapo prison, see Nathan Stoltzfus, *Resistance of the Heart: Intermarriage and the Rosenstrasse Protest in Nazi Germany*. New York: W.W. Norton, 1996.
30. Klukowski, op. cit., p. 199.

31. Szwajger, *I Remember Nothing More*, pp. 46–7.
32. Abraham Lewin, *A Cup of Tears: A Diary of the Warsaw Ghetto*. Ed. by Antony Polonsky. Trans. by Christopher Hutton. Oxford: Basil Blackwell, 1988, p. 154. An interesting account of resistance and escape from Sobibor can be found in Thomas Toivi Blatt, *From the Ashes of Sobibor: A Story of Survival*. Foreword by Christopher R. Browning. Evanston, IL: Northwestern University Press, 1997; and Miriam Novitch, *Sobibor, Martyrdom and Revolt: Documents and Testimonies*. Preface by Leon Poliakov. New York: Holocaust Library, 1980.
33. Alexander B. Rossino, 'Destructive Impulses: German Soldiers and the Conquest of Poland'. *Holocaust and Genocide Studies*, Vol. 11, No. 3, Winter 1997, pp. 351–65, p. 355. See also Shraga Feivel Bielawski, *The Last Jew From Wegrow: The Memoirs of a Survivor of the Step-By-Step Genocide in Poland*. SS leader Globocnik, regarding resettlement in Lublin, observes: 'German blood has been saved [and] ... this blood can be made use of for the future strengthening and security of our entire *Volkstrum*.' Quoted in Isabel Heinemann, 'Another Type of Perpetrator: The S.S. Racial Experts and Forced Population Movements in the Occupied Regions'. *Holocaust and Genocide Studies*, Vol. 15, No. 3, Winter 2001; see also Jacob Apenszlak, ed., *The Black Book of Polish Jewry*. New York: The American Federation for Polish Jews, 1943.
34. Yitzhok Rudashevski, *The Diary of the Vilna Ghetto: June 1941–April 1943*. Trans. by Percy Matenko. Jerusalem: Ghetto Fighters House, 1973, pp. 36, 38.
35. Rusashevski, *Diary of the Vilna Ghetto*, pp. 39–40.
36. Ibid., pp. 41–2.
37. Ibid., p. 43.
38. Ibid., p. 46.
39. Ibid., p. 56.
40. Ibid., p. 57.
41. Ibid., pp. 91–2.
42. Ibid., pp. 101, 106.
43. Ibid., p. 99.
44. Ibid., p. 117.
45. Ibid.
46. Ibid.
47. Ibid., p. 90.
48. March 27 and 28, 1944: Ephraim Oshry, *The Annihilation of Lithuanian Jewry*. Trans. by Y. Leiman. New York: The Judaica Press, 1995, p. 123.
49. Ibid., pp. 124–5.
50. Lewin, *A Cup of Tears*, pp. 28–9.
51. Ibid., p. 35.
52. See Tory, *Surviving the Holocaust*, p. 176.
53. Kowalski, pp. 300–1.
54. Ibid., p. 274.
55. Ibid., p. 228.
56. Ibid., p. 202.

57. Ibid.
58. Ibid., p. 201.
59. Ibid., p. 198.
60. Ibid., p. 128.
61. Ibid., p. 123.
62. Ibid.
63. Ibid., p. 82.
64. Yitzhak Arad, *Ghetto in Flames: The Struggle and Destruction of the Jews in Vilna in the Holocaust*. Jerusalem: Yad vaShem, Martyrs' and Heroes' Remembrance Authority, 1980, p. 178.
65. Ibid., p. 254.
66. Ibid., p. 400.
67. Ibid., p. 417.
68. Ibid., p. 418.
69. Newsletter, 'Poland Fights'. August 15, 1951, in *Black Book*, p. 46.
70. *Black Book*, pp. 46–9.
71. Ibid., p. 59. Tosia Bialer writes in *Collier's* (February 20, 1943): 'There were tens of thousands of families who could not afford black-market prices and had to depend on the rationed goods for subsistence. Slow victims of undernourishment, these. Their teeth decayed and fell out, hair and nails refused to grow, their eyes became great sunken hollows in fleshless faces, and their stomachs were repulsively bloated. These miserable travesties of human beings picked up what they could find in the streets and in garbage piles, consuming the rest of their strength in the awful fight against real starvation' (p. 55).
72. Ibid., p. 59. She saw undertakers' vans scour the city all day collecting emaciated bodies from which all clothing had been stripped, and transporting them to cemeteries where they were buried three-deep in mass graves. 'Orphaned children with spindly legs and famine-bloated bodies roamed the ghetto streets begging for food, but nobody had much to give them. It had become almost commonplace, she said, to see children and adults drop dead from starvation in the streets. Scores of people committed suicide every day… . The grave is a ditch 30 yards by 20 yards containing naked bodies of men, women and children' (p. 59).
73. Ibid., p. 64.
74. Ibid., pp. 198, 199.
75. Ringelblum, *Polish Jewish Relations during the Second World War*, p. 157.
76. Ibid., p. xxvii.
77. Ibid., pp. 159–60. A helpful discussion and analysis of Jewish leadership during the Holocaust is Raul Hilberg, *Patterns of Jewish Leadership in Nazi Europe, 1933–1945*. New York: Holmes and Meier, 1995.

3 The Moral Position of Violence: Bielski Survivors

1. Nechama Tec, *Defiance: The Bielski Partisans*. New York: Oxford University Press, 1993, p. 147; see also Lester Samuel Eckman and Chaim Lazar, *The*

176 Notes

    Jewish Resistance: The History of the Jewish Partisans in Lithuania and White Russia during the Nazi Occupation, 1940–1945. New York: Shengold, 1977. For detailed histories of the partisans and their activities, see Allan Levin, *Fugitives of the Forest: The Heroic Story of Jewish Resistance and Survival during the Second World War.* New York: Stoddart, 1998. For an analysis of the complex and often tense relationship between the Jewish and Soviet partisans, see Hersh Smolar, *The Minsk Ghetto: Soviet-Jewish Partisans against the Nazis.* New York: Holocaust Library, 1989; Jack Kagan, *Surviving the Holocaust with Russian Jewish Partisans.* New York: Vallentine Mitchell and Co., 1998. Kagan gives a harrowing account of the tunnel escape from Novogrudek; for another compelling narrative, see Liza Ettinger, *From the Lida Ghetto to the Bielski Partisans.* Memoir in the United States Holocaust Memorial Museum, archives RG-02.133, 1984.
2. Tec, op. cit. See also Shalom Cholawski, *The Jews of Belorussia during World War II.* New York: Harwood Academic Publishers, 1998.
3. Ibid., p. 151. For a graphic account of the kind of punishment meted out by the partisans, see John A. Armstrong, ed., *Soviet Partisans in World War II.* Madison: University of Wisconsin Press, 1964.
4. Ibid., p. 45.
5. Ibid., p. 47. For an excellent account of the Soviet partisan movement, see Kenneth Slepyan, 'The Soviet Partisan Movement and the Holocaust', *Holocaust and Genocide Studies*, vol. 14, no. 1, Spring 2000.
6. Ibid., p. 83.
7. Ibid., p. 129.
8. Ibid., p. 136.
9. Ibid., p. 139.
10. Also helpful in explaining the harsh conditions of the Nazi Occupation of Novogrudek was a taped interview with survivor Rae Kushner (United States Holocaust Memorial Museum archives, RG 50.002*0015). Killings throughout the region are described in chilling detail in Martin Dean, *Collaboration in the Holocaust: Crimes of the Local Police in Belorussia and Ukraine, 1941–44.* New York: St. Martin's Press, 2000. Inside Germany persecution and roundups of Jews by the Gestapo elicited little reaction. Eric A. Johnson (in *Nazi Terror: The Gestapo, Jews and Ordinary Germans.* New York: Basic Books, 1999) writes: 'Most of the ordinary German population supported the Nazi regime, did not perceive the Gestapo as all-powerful or even as terribly threatening to them personally, and enjoyed considerable room to express frustration and disapproval arising out of minor disagreements with the Nazi state and its leadership' (p. 262).

4 The Moral Goodness of Violence: Necessity in the Forests

1. Shalom Cholawski, *Soldiers from the Ghetto.* San Diego: A.S. Barnes and Company, Inc., 1980, p. 95. See also Shalom Yoram, *The Defiant: A True*

Story. Trans. by Varda Yoram. New York: St. Martin's Press, 1996, p. 131. Cf. Zvi Gitelman, ed., *Bitter Legacy: Confronting the Holocaust in the USSR*. Bloomington: Indiana University Press, 1997; Jan T. Gross, *Revolution from Abroad: The Soviet Conquest of Poland's Eastern Ukraine and Western Belorussia*. Princeton, NJ: Princeton University Press, 1988. Not widely understood was the extent of the *Reichsbahn* [German railway] contributions to the machinery and process facilitating mass murder, not only in the camps but in the conquered territories of the East. For a recent exploration of their role, see Alfred C. Mierzejewski, 'A Public Enterprise in the Service of Mass Murder: the Deutsche Reichsbahn and the Holocaust', *Holocaust and Genocide Studies*, vol. 15, no. 1, Spring 2001.
2. Cholawski, *Soldiers from the Ghetto*, p. 114. See also Jack Nusan, ed., *Jewish Partisans: A Documentary of Jewish Resistance in the Soviet Union during World War II*. Trans. from Hebrew by the Magal Translation Institute, Ltd., based on Russian, Polish and Yiddish sources. Washington, D.C.: University Press of America, 1982.
3. Frantz Fanon, *The Wretched of the Earth*. Trans. by Constance Farrington. New York: Grove-Atlantic, 1988. For a provocative analysis of the negative side of this proposition and the genocidal consequences of violence, see James Waller, *Becoming Evil: How Ordinary People Commit Genocide and Mass Killing*. New York: Oxford University Press, 2002. Waller is looking at the reasons behind genocide, not at the violence that meets or faces the genocidal killer. Fanon does not explore the extent to which redemptive violence (as opposed to the annihilatory violence of the indiscriminate mass murderer) might degenerate into genocidal violence, although his final chapter on mental disorders suggests that wherever violence touches, injury follows. This last chapter, an extraordinary psychiatric investigation of the psychological consequences of violence, shows ambivalence in Fanon's argument not present in the earlier discussion.
4. Faye Schulman, *A Partisan's Memoir: Women of the Holocaust*. Toronto: Second Story Books, 1995, p. 175. See also Lester Samuel Eckman and Chaim Lazar, *The Jewish Resistance: The History of the Jewish Partisans in Lithuania and White Russia during the Nazi Occupation, 1940–1945*. New York: Shengold, 1977; and Dov Cohen and Jack Kagan, *Surviving the Holocaust with the Russian Jewish Partisans*. Portland, OR: Vallentine Mitchell, 1998.
5. Sutin and Sutin, *Jack and Rochelle*, pp. 143, 142. See also Zoe Szner, ed., *Extermination and Resistance: Historical Records and Source Material*. Haifa: Ghetto Fighters House, 1958. 'Fighting back' often had as much to do with the enthusiasm of local leaders and militias for killing as with the Germans. See Wendy Lower's interesting study ('"Anticipatory Obedience" and the Implementation of the Holocaust in the Ukraine: A Case Study of Central and Peripheral Forces in the Generalbezirk Zhytomyr, 1941–1944', *Holocaust and Genocide Studies*, vol. 16, no. 1, Spring 2002, pp. 8–22). Lower speaks of the extent of willed participation

amongst collaborators and indigenous populations. 'as of early 1942 even verbal orders were deemed unnecessary for authorizing the murderous "mopping-up" actions against Jews in hiding. Thus more local leaders learned what was expected of them, and fewer needed explicit orders to do it. The commissars and regional police forces did not carry out the Nazi goal of genocide in a banal fashion ...often encouraging sadistic methods that exceeded the expectations of their superiors' (p. 8).

6. Isaiah Trunk, *Jewish Responses to Nazi Persecution: Collection on Individual Behavior in Extremis*. New York: Stein and Day, 1979, p. 250. For a comprehensive review of the statistics, see Levin, *Fighting Back*, pp. 179–203.
7. Trunk, *Jewish Responses*, pp. 246, 250.
8. Ibid., p. 303. For an account written shortly after the war, see Marie Syrkin, *Blessed is the Match: The Story of Jewish Resistance*. Philadelphia: Jewish Publication Society of America, 1948.
9. Fanon, *The Wretched of the Earth*, p. 37.
10. Ibid., p. 69.
11. Ibid., p. 73.
12. Ibid. To understand this notion in the context of Jewish resistance, see John Sack, *An Eye for an Eye: The Untold Story of Jewish Revenge against Germans*. New York: Basic Books, 1993.
13. Ibid., p. 72.
14. Ibid. For an idea of this sense of self in the Jewish resistor, see Hanna Krall, *Shielding the Flame: An Intimate Conversation with Dr. Marek Edelman, the Surviving Leader of the Wars*aw *Ghetto Uprising*. New York: Henry Holt, 1986. The recognition of Jewish resistance and its significance appeared even before the war's end. See Mac Davis, *Jews Fight Too*. New York: Jordon Publishing Co., 1945.
15. Ibid., p. 46.
16. Ibid.
17. Ibid.
18. Ibid., p. 45.
19. Ibid., p. 38.
20. Ibid., p. 37.
21. Ibid., p. 63.
22. Ibid., p. 73.
23. Ibid., p. 111.
24. See Cohen and Kagan, *Surviving the Holocaust with Russian Jewish Partisans*, p. 91. Also *Forests of Valor: Following in the Footsteps of Jewish Partisans, U.S.S.R.* (video), April, 1989. Producer, Issy Avron; written and directed by Zvi Godel. Israel Educational Television. Teaneck, NJ: Ergo Media, 1996; Issac Kowalski, ed., *Anthology on Armed Jewish Resistance, 1939–1945*, 4 vols. Brooklyn: Jewish Combatants Publishers House, 1986.
25. Levin, *Fighting Back*: 'the total number of people [in Lithuania] who participated in the active fight against the Nazis comes to over 2,000, that is approximately 5 percent of the 40,000 Jews who remained in Lithuania at the beginning of 1942 after the previous mass liquidation'

(p. 175). For an extraordinary video of survivors of the Vilna underground, see *Partisans of Vilna*. Ciesia Foundation: directed and edited by Josh Waletzky; produced by Aviva Kempner. Washington, D.C.: Euro-American Home Video, 1987. See also Herman Kruk, *The Last Days of the Jerusalem of Lithuania: Chronicles from the Vilna Ghetto and the Camps, 1939–1944*. Ed. and intro. by Benjamin Harshaw. Trans. by Barbara Harshaw. New York: YIVO Institute for Jewish Research, 2002.
26. Yehuda Bauer, *The Jewish Emergence from Powerlessness*. Toronto: University of Toronto Press, 1979, p. 28. See also Leon Kahn, *No Time to Mourn: A True Story of a Jewish Partisan Fighter*. Vancouver: Laurelton Press, 1978. See also *Yehude Yaar* [Forest Jews], *Narratives of Jewish Partisans of White Russia, Tuvia and Zus Bielski, Lilka and Sonia Bielski and Abraham Viner*, as told to Ben Dor (Tel Aviv: Am Oved, 1946). This extraordinary set of interviews has been translated into English in a private publication by R. Goodman. It is not available to the general public.
27. Yitzhak Arad, *The Partisan: From the Valley of Death to Mount Zion*. New York: Holocaust Library, 1979, p. 81.
28. See Adina Blady Szwajger, *I Remember Nothing More: The Warsaw Children's Hospital and the Jewish Resistance*. Trans. by Tasja Darowska and Danusia Stok. New York: Pantheon Books, 1990, p. 80. See also Philip Friedman, 'Jewish Resistance to Nazism'. In Ada J. Friedman, ed., *Roads to Extinction: Essays on the Holocaust*. Philadelphia: the Jewish Publication Society, 1980, pp. 387–408.
29. Yitzhak Arad, *Ghetto in Flames: The Struggle and Destruction of the Jews of Vilna in the Holocaust*. Jerusalem: Yad vaShem, 1980, pp. 411–12.
30. Shalom Yoram, *The Defiant: A True Story*. Trans. by Varda Yoram. New York: St. Martin's Press, 1996, p. 108. See also Shmuel Krakowski, *The War of the Doomed: Jewish Armed Resistance in Poland, 1942–1944*. Trans. by Orah Blaustein. New York: Holmes and Meier, 1984.
31. Ibid., p. 109.
32. Ibid., pp. 172–3.
33. Ibid., p. 183.
34. Arad, *The Partisan*, pp. 119–20.
35. Ibid., p. 95.
36. Nahum Kohn and Howard Roiter, *A Voice from the Forest*. New York: Holocaust Library, 1980, p. 15.
37. Ibid., p. 32.
38. Ibid., pp. 33, 37.
39. Ibid., p. 95.
40. Ibid., p. 246; see the extraordinary interview of survivors, including the Bielski brothers, of this unit. *The Bielski Brothers: the Unknown Partisans*. SOMA Productions; directed by Arun Kumar; written and produced by David Herman. Princeton, NJ: Films for the Humanities and Sciences, 1996.
41. See Kowalski, *Anthology on Armed Jewish Resistance*; 'Partisans'. In *Encyclopedia of the Holocaust*. London: Macmillan, 1990, vol. 4; Yisrael

Gutman, *Fighting Among the Ruins: the Story of Jewish Heroism During World War II*. Washington, D.C.: B'nai B'rith Books, 1988; Gilles Lambert, *Operation Hazalah: How Young Zionists Rescued Thousands of Hungarian Jews in the Nazi Occupation*. Trans. by Robert Bullen and Rosette Letellier. New York: Bobbs-Merrill, 1974.
42. I use Rousseauian in the sense of the strong community Rousseau describes in his *Social Contract*, particularly the unifying and adhesive concept of the General Will. For Rousseau what is common to the community precedes all other special interests; also what is 'common' possesses an exclusivity in terms of allegiance and commitment. Jean-Jacques Rousseau, *The Social Contract and Discourses*. Trans. by G.D.H. Cole. New York: E.P. Dutton, 1951.
43. Sigmund Freud, *The Future of an Illusion*. Trans. by James Strachey. New York: Norton, 1975.

5 Spiritual Resistance: Understanding its Meaning

1. Gertrude Hirshler and Shimon Zucker, eds., *The Unconquerable Spirit: Vignettes of the Religious Spirit the Nazis Could Not Destroy*. Trans. by Gertrude Hirshler. New York: Mensora Publications, 1981; cf. Y. Gottfarstein, 'Kiddush ha Shem in the Holocaust Period'. In *Jewish Resistance during the Holocaust*; Peter Haas, *Responsa: Literary History of a Rabbinic Genre*. Atlanta: Scholars Press, 1996.
2. See Yaffa Eliach, *Hasidic Tales of the Holocaust*. New York: Oxford University Press, 1982, Chapter 3. The emotional and psychological demands placed on rabbis were enormous. See Saul Esh, 'The Dignity of the Destroyed; Towards a Definition of the Period of the Holocaust'. *Jewish Magazine*, vol. XI, no. 2, Spring 1962.
3. Underground literature, with rare exceptions, never targets rabbinical inaction. Similarly in diaries like David Rudashevski's or Morris Berg's *Warsaw Ghetto Diary* (New York: L.B. Fischer Co., 1945) there is an absence of critical comments on rabbis. Cf. Zelig Kalmanovitch, *A Diary of the Nazi Ghetto in Vilna*. YIVO. Annual of Jewish Social Studies. Vol. III. New York: Yiddish Scientific Institute, 1953.
4. Joseph Rudavsky, *To Live with Hope: To Die with Dignity*. Lanham, MD: University Press of America, 1987, p. 157.
5. Cf. Shimon Huberband, *Kiddush Hashem: Jewish Religious and Cultural Life in Poland during the Holocaust*. Trans. by David E. Fishman. New York: Yeshiva University Press, 1987. This is an extraordinary, contemporaneous account of brutalization and endeavor by Polish Jews during the Nazi occupation. Yet, with all its seriousness and tragedy, Huberband includes a section on 'Wartime Folklore,' what we would call jokes. An example of this kind of gallows humor is the following: 'A teacher asks his pupil, "Tell me Moishe, what would you like to be if you were Hitler's son?" "An orphan," the pupil answers' (p. 113).

6. Cf. Robert Kirschner, trans., *Rabbinic Responsa of the Holocaust Era*. New York: Schocken Books, 1985; David Kraemer, *Responses to Suffering in Classical Rabbinic Literature*. New York: Oxford University Press, 1995.
7. See Pesach Schindler, *Hasidic Responses to the Holocaust in the Light of Hasidic Thought*. Hoboken, N.J., 1990. *Kiddush haShem* 'was the opportunity to counter the Satanic and the impure with the elements that were the most difficult to destroy – the spiritual and divine in human existence' (p. 117).
8. But spiritual resistance did have tremendous efficacy in the following sense: spiritual resistance meant having the choice to go 'to one's death degraded and dejected as opposed to confronting [death] with an inner peace, nobility, upright stance, without lament and cringing to the enemy.' Meir Dvorzeski, quoted in Schindler, *Hasidic Responses*, p. 61.
9. See Menashe Unger, *Sefer Kedoshim: Rebeim of Kiddush Hashem* [the Book of Martyrs: Hasidic Rabbis as Martyrs During the Holocaust]. New York: Shul Singer, 1967; see also *Rehilat Haharedim Collection*. Files on the activities of the Habad movement (Lubaviticher Hasidim) during the Nazi Occupation of France. In YIVO archives. New York.
10. Cf. Max Kaddushim, *The Rabbinic Mind*. New York: Jewish Theological Seminary, 1952.
11. Cf. Jacob Robinson and Philip Friedman, *Guide to Jewish History under Nazi Impact: Joint Documentary Projects*, Bibliographical Series, No. 1. New York: YadVashem and YIVO, 1960.
12. Primo Levi, *The Drowned and the Saved*. Trans. by Raymond Rosenthal. New York: Random House, 1989.
13. Shalom Cholawski, *Soldiers from the Ghetto*. San Diego: A.S. Barnes & Co., Inc., 1980, p. 120. See also Hermann Wygoda, ed., *In the Shadow of the Swastika*. Foreword by Michael Berenbaum. Urbana, IL: University of Illinois Press, 1998; (video), *Resistance*. Produced and directed by C.J. Pressma. New York: Cinema Guild, 1982.
14. Cf. Joseph Gar, *Bibliography of Articles on the Catastrophe and Heroism in Yiddish Periodicals*. Joint Documentary Projects, No. 9. New York: Yadvashem and YIVO, 1966; *Bibliography of Articles on the Catastrophe and Heroism in North American Yiddish Periodicals*. Joint Documentary Projects, No. 10. New York: YadVashem and YIVO, 1969.
15. Isaac C. Avigdor, *From Prison to Pulpit: Sermons and Stories*. Hartford, CT: Hanav Publishers, 1975, p. 215; cf. Menahem Brayer, 'The Hasidic Rebbes of Romania, Hungary and their Relationship to Eretz Yisrael'. In *Hasidut Verzion* [Hasidism and Zion]. Ed. by Simon Federbush. New York: Moriah, 1963.
16. For an insightful analysis of Hasidic practices and ethics see Martin Buber, *The Origin and Meaning of Hasidism*. New York: Harper and Row, 1960; Gershon G. Scholem, *Major Trends in Jewish Mysticism*. New York: Schocken Books, 1954; Zalman Schuchter-Shalomi, *Wrapped in a Holy Flame: Teachings and Tales of the Hasidic Masters*. San Francisco: Jossey-Bass, 2003; Tzvi Rabinowicz, ed., *The Encyclopedia of Hasidism*. Northvale, NJ: Jason Aronson, 1996.

17. Avigdor, *From Prison to Pulpit*, p. 100. A Hasidic Jew in a small Polish ghetto describes German atrocities: 'What can they do to me. They can take my body – but not my soul! Over my soul they have no dominion! Their dominion is only in this world. Here they are the mighty ones. All right. But in the world to come their strength is no more.' Irving J. Rosenbaum, *The Holocaust and Halakhah*. Jerusalem: Ktav Publishing House, Inc., 1976, pp. 35–6.
18. Nehemia Polen, *The Holy Fire: The Teachings of Rabbi Kalonymus Kalman Shapira, the Rebbe of the Warsaw Ghetto*. Northvale, NJ: Jason Aronson, Inc., 1994, p. 100. For a fascinating analysis of the connections between the rabbinic sermon and the imagination see Lewis M. Barth, 'Literary Imagination and the Rabbinic Sermon'. Barth argues that the rabbinical sermon moves to establish a hope based on the restoration of a 'social order' that will be testament to God's will, since God's intent is to restore 'judges and leaders embodying qualities of ancient heroes' (p. 31). Whether this was the 'hope' of Rabbi Shapira, we don't know; but there is evidence in his sermons that he believed this might happen. From *Proceedings of the Seventh World Congress of Jewish Studies: Studies in the Talmud, Halacha and Midrash*. Held at the Hebrew University of Jerusalem, August 7–14, 1977. Jerusalem: Magnes Press, 1981.
19. Ibid., pp. 102–3. Some testimony suggests that even in Auschwitz, the signs and symbols of spiritual identity took on an extraordinary significance. 'The burning candle kindled in our hearts new hope for the future and strengthened our trust in the "rock of our salvation".' From the writing of Rabbi Sinai Adler, quoted in Rosenbaum, *The Holocaust and Halakhah*, p. 118.
20. Ibid., p. 103.
21. Ibid., p. 119.
22. Ibid., p. 120.
23. Ibid., p. 130.
24. Ibid., pp. 131–2. The rabbis continually faced serious questions; for example, 'Aaron Rapoport a Hasid of the Ostrowzer Rebbe, Rabbi Yechezkel Halstik, confronts his rebbe hiding under miserable conditions and asks. "Is this the Torah and this its reward? What is happening here?" The rebbe answers that man may be able to probe the soul of his fellow man but not the ways of God. "I was very bitter and refused to ask any more questions"' (cited Schindler, *Hasidic Responses*, p. 24).
25. Ibid., p. 132.
26. One of the most extraordinary diaries is that of Chaim A. Kaplan, a detailed account of both physical and psychological disintegration. Chaim A. Kaplan, *Scroll of Agony: The Warsaw Diary of Chaim A. Kaplan*. Trans. by Abraham I. Katsch. New York: Macmillan, 1965. Or cf. the following from Hillel Seidman's *Warsaw Ghetto Diary*. 'People in the ghetto whose normal moral resistance had been low descended to an even lower level [in the ghetto].' Yet there were 'those of

high moral integrity [who] rose to even greater heights' (cited in Schindler, *Hasidic Responses*, p. 120).
27. Cf. Gotz Aly, Peter Chroust and Christian Pross, *Cleansing the Fatherland: Nazi Medicine and Racial Hygiene*. Trans. Belinda Cooper. Baltimore: Johns Hopkins Press, 1994; Christopher R. Browning, *Ordinary Men: Reserve Police Battalion 101 and the Final Solution in Poland*. New York: Harper Perennial, 1992; Daniel Jonah Goldhagen, *Hitler's Willing Executioners: Ordinary Germans and the Holocaust*. New York: Knopf, 1996; Benno Muller Hill, *Marvelous Science: Elimination by Scientific Selection of Jews, Gypsies and Others, Germany 1933–1945*. Trans. by George R. Fraser. New York: Oxford University Press, 1988.

6 Condemned Spirit and the Moral Arguments of Faith

1. Moses Maimonides, *Treatise on Resurrection*. Trans. by Fred Kosner. New York: Ktav Publishing House, Inc., 1982, p. 25.
2. Ibid., pp. 25–7.
3. Ibid., p. 22.
4. Shimon Huberband, *Kiddush haShem: Jewish Religious and Cultural Life in Poland during the Holocaust*. Hoboken, NJ: Ktav Publishing House, Inc., 1987, p. 123.
5. For an account of daily acts of ritual protest, see ibid., pp. 175–239.
6. Ibid., p. 199.
7. Ibid., p. 201.
8. Ibid., pp. 240–1. Conditions inside the ghetto had deteriorated to such an extent that for many death and poverty became as ordinary as the cobblestones on the street. Huberband captures something of this commonplace property of death in the following observation: 'Near a display window of a large concern of various pastries, wines, liquors, grapes and other delicacies, there lay the dead body of a Jew, some thirty-odd years old. The dead body was totally naked. It was a truly ironic scene; near a display window of such delicacies lay the dead body of a Jew who had died of starvation. Nonetheless, this did not prevent the dressed-up ladies from walking across the dead body to enter the store and then leave [the store] with packages of goodies which, if only a fraction of the contents of those packages had been given to the hungry, this Jew would not have died of starvation' (p. 240).
9. Ibid., pp. 241–2.
10. Ibid., pp. 264–5.
11. Irving J. Rosenbaum, *Holocaust and Halakhah*. Jerusalem: Ktav Publishing House, Inc., 1976, pp. 35–6.
12. Ibid., p. 36.
13. Ibid., p. 116. In Kovno and Warsaw, some contemporary observers argue that the suicide rate was quite low (cf. Chaim Kaplan, *The Warsaw Diary of Chaim Kaplan*, p. 131). In Lodz, however, it was quite high. Cf. for a frightening account of the daily rise of suicides, see Lucjan Dobroszycki

and Danuta Dablowska, eds. *The Chronicle of the Lodz Ghetto.* Trans. Richard Lourie. New Haven: Yale University Press, 1984. It is impossible to gauge exact numbers, but a reasonable conclusion would be that while suicide rates varied from ghetto to ghetto, it was still the case that suicides rose dramatically during the Holocaust, and that rabbis frequently found themselves in the position of passing moral judgments on acts of suicide. The law could both justify and condemn suicide. See Rosenbaum, *Holocaust and Halakhah,* pp. 161n, 162n.
14. Janusz Korczak, *The Ghetto Years: 1939-1941.* Trans. by Jerzy Bachrack and B. Krzywicka [Vedder]. New York: Holocaust Library, 1980, p. 208.
15. Ibid., pp. 208, 64.
16. Nehemia Polen, *The Holy Fire: The Teachings of Rabbi Kalonymus Kalman Shapira, the Rebbe of the Warsaw Ghetto.* Northvale, NJ: Jason Aronson, Inc., 1994, p. 19.
17. Ibid., p. 25.
18. Ibid., p. 26.
19. Ibid., pp. 30, 31.
20. Ibid., p. 35.
21. Ibid., pp. 37-8.
22. Ibid., p. 39. But this passage does not imply rabbinical quiescence. Shapira was active in his efforts to bring both spiritual and physical assistance to the suffering. This was the case with the vast majority of the rabbis. See Pesach Schindler, *Hasidic Responses to the Holocaust in the Light of Hasidic Thought.* Hoboken, NJ: Ktav Publishing House, Inc., 1990, p. 76.
23. Ibid., p. 40.
24. Ibid., p. 45.
25. Ibid., p. 65.
26. Robert Kirschner, *Rabbinic Responsa of the Holocaust Era.* Trans. by Robert Kirschner. New York: Schocken Books, 1985, p. 69.
27. Huberband, *Kiddush haShem,* p. 123. For Polish participation and its intensity in the roundup and persecution of Jews, see Gross, *Neighbors,* an account of the slaughter of Jews in a small Polish village.
28. Rabbi Kalonymus Kalman Shapira, *A Student's Obligation: Advice from the Rebbe of the Warsaw Ghetto.* Trans. by Michael Odenheimer. Northvale, NJ: Jason Aronson, Inc., 1991, p. 68.
29. Kirschner, *Rabbinic Responsa,* pp. 113ff.
30. H. Krall, *Shielding the Flame: An Intimate Conversation with Dr. Marek Edelman, the Last Surviving Leader of the Warsaw Ghetto Uprising.* Trans. by J. Stansinska and C. Weschler. New York: Henry Holt and Company, 1986, p. 47.
31. Cf. Daniel J. Goldhagen, *Hitler's Willing Executioners: Ordinary Germans and the Holocaust.* New York: Alfred A. Knopf, 1996, p. 311.
32. Ibid., p. 316.
33. The resistance had no patience with this argument: Vernon: 'You know, the Germans could kill souls ...You needed to figure out a way to protect yourself, and it wasn't easy!' For a fascinating account of the

strategy of protection, see Yitzhak Arad, *The Partisan: From the Valley of Death to Mt. Zion*. New York: Holocaust Library, 1979.
34. A. Adelson and R. Lapides, eds., *Lodz Ghetto: Inside a Community under Siege*. New York: Penguin Books, 1991, p. 161. Nutritional need for laboring persons generally lies in excess of 3,000 calories a day; the average daily caloric intake for ghetto inhabitants, however, consisted of 700–900 calories a day. In Lodz, for example, workers and the few children and elderly that remained after 1943 faced the everyday possibility of death from minor illnesses due to insufficient rations shared in families. But even with the Nazi policy of killing through work, the Jewish population of Lodz managed to produce munitions, telecommunications equipment, uniforms, boots, lingeries, temporary housing and carpets. Hans Biebow, the commandant of the Lodz ghetto, became a millionaire through his appropriation of ghetto profits. He was executed shortly after liberation. Lodz survived the longest of the ghettos established in Eastern Europe to concentrate Jews. The ghetto lay in a section of Poland that had, after the 1939 invasion, been annexed by Germany to the greater Reich. And some disagreement over its fate occurred in higher German administrative units. Albert Speer and Heinrich Himmler, for example, disagreed over the ghetto workshops' economic benefits. Speer found it useful for the production of munitions. Eventually, Himmler's genocidal argument overruled Speer's strictly economic focus; and in 1944 Jews remaining in Lodz perished in Auschwitz. Of the almost 200,000 initially held in the sealed ghetto, 80,000 were alive before deportation to Auschwitz in the summer of 1944. Between the establishment of the ghetto in 1940 and its destruction in 1944, 60,000 died from starvation, cold, disease, hanging or suicide.
35. Primo Levi, *The Drowned and the Saved*. Trans. by Raymond Rosenthal. New York: Random House, 1989, pp. 179–80.
36. Polen, *The Holy Fire*, p. 72.
37. Lawrence L. Langer, *Admitting the Holocaust: Collected Essays*. New York: Oxford University Press, 1995, p. 49.
38. Adelson and Lapides, *Lodz Ghetto*, pp. 420–1.
39. Polen, *The Holy Fire*, p. 81.
40. Huberband, *Kiddush haShem*, p. 334.
41. Adelson and Lapides, *Lodz Ghetto*, p. 349.
42. Ibid., p. 355.
43. Polen, *The Holy Fire*, p. 83.
44. Ibid.
45. Ibid.
46. Ibid., p. 84.
47. Ibid., p. 86.
48. Ibid., p. 88.
49. Ibid., p. 89.
50. Ibid., pp. 89–90.
51. Ibid., pp. 90–1.

7 The Silence of Faith Facing the Emptied-out Self

1. Lawrence L. Langer, *Holocaust Testimonies: The Ruins of Memory*. New Haven, CT: Yale University Press, 1991, p. 25.
2. Ibid., p. 12.
3. Ibid., pp. 15–16.
4. Ibid., p. 33.
5. Ibid., p. 48.
6. Ibid., p. 49.
7. Ibid., p. 59.
8. Ibid., pp. 63–4.
9. Ibid., p. 65.
10. Ibid., p. 68.
11. Ibid., p. 71.
12. Ibid., p. 84.
13. Ibid., p. 86.
14. Ibid., p. 96.
15. Ibid., p. 119.
16. Ibid., p. 131.
17. Cf. D.W. Winnicott, *Collected Papers*. London: Tavistock, 1958; see also this concept in R.D. Laing, *The Divided Self*. New York: Penguin, 1978.
18. Langer, *Holocaust Testimonies*, p. 136.
19. Ibid., p. 140.
20. Ibid., pp. 148–9.
21. Ibid., p. 149.
22. Ibid., p. 150. Pesach Schindler's meticulously researched book on Hasidic responses to oppression and violence suggests that spiritual resistance was a far more intense and difficult experience and should not be understood as a 'veneer of respectability' for impossible moral situations.
23. Ibid., pp. 165–6.
24. Ibid., p. 165.
25. Ibid., p. 168.
26. Ibid., p. 176.
27. Ibid., p. 177.
28. Ibid., p. 178.
29. Ibid., p. 180.
30. Ibid., p. 180.
31. Shimon Zuker, *The Unconquerable Spirit: Vignettes of the Jewish Religious Spirit the Nazis Could Not Destroy*. Trans. by Gertrude Hirschler. New York: Mesorah Publications, p. 28. See also *Spiritual Resistance: Art from Concentration Camps, 1940–1945: a Selection of Drawings and Paintings from the Collection of Kibbutz Lochamei HaGhettaot Israel*. With essays by Miriam Novitch, Lucy Dawidowicz, Tom L. Freudenheim. New York: Union of American Hebrew Congregations, 1981.
32. Quoted in Steven L. Jacobs, ed., *Studies in the Shoah. Contemporary Jewish Religious Responses to the Shoah*. Lanham, MD: University Press of America, 1993. Vol. V, p. 142.

33. Zuker, *The Unconquerable Spirit*, p. 28.
34. Ibid., p. 129.
35. Nachman Blumental, 'Magical Thinking among the Jews during the Nazi Occupation'. In Nathan Eck and Arieh Leon Kabovy, eds., *Yad VaShem Studies on the European Jewish Catastrophe and Resistance*. Vol. 5, 1963, p. 226.
36. Ibid., p. 231.
37. Zuker, *The Unconquerable Spirit*, p. 7. See also Rabbi Ephraim Oshry, *The Annihilation of Lithuanian Jewry*. Trans. by Y. Leiman. *The Yiddish Book, Churbvan Lila*. New York: The Judaica Press, Inc., 1995.
38. Quoted in H.S. Zimmels, *The Echo of the Nazi Holocaust in Rabbinical Literature*. London: Ktav Publishing House, 1977, p. 58.
39. Ibid., p. 58.
40. Ibid., p. 64.

8 Law and Spirit in Terrible Times

1. Cf. Irving J. Rosenbaum, *Holocaust and Halakhah*. Jerusalem: Ktav Publishing House, Inc., 1976.
2. Ibid., p. 4.
3. Ibid., p. 5.
4. Ibid., p. 65.
5. Ibid., pp. 5–6.
6. Ithamar Gruenwald, *Apocalyptic and Merkavah Mysticism*. Leiden: E.J. Brill, 1980, p. 153.
7. Pesach Schindler, *Hasidic Responses to the Holocaust in the Light of Hasidic Thought*. Hoboken, NJ: Ktav Publishing House, Inc., p. 64.
8. Ibid., p. 65.
9. Ibid.
10. Ibid., pp. 164–5.
11. Ibid., p. 49.
12. Ibid., pp. 49–58.
13. Ibid., p. 56.
14. Ibid.
15. Ibid., p. 21.
16. Ibid., p. 27.
17. Ibid., p. 35.
18. Ibid., p. 36.
19. Ibid., p. 38.
20. Berkovits, *Faith after the Holocaust*, p. 69.
21. Ibid.
22. Ibid., p. 81.
23. A. Adelson and R. Lapides, *Lodz Ghetto: Inside a Community under Siege*. New York: Penguin Books, 1991, pp. 348–9.
24. Joseph Horn, *Mark it with a Stone*. New York: Barricade Books, 1996, p. 68.

25. Ibid., p. 71.
26. Ibid., p. 77.
27. David Kraemer, *Responses to Suffering in Classical Rabbinic Literature*. New York: Oxford University Press, 1995, p. 94.
28. Ibid., p. 93.
29. Ibid., p. 82.
30. Horn, *Mark it with a Stone*, p. 81.
31. Ibid., p. 190.

Bibliography

Adelson, A. and R. Lapides. 1991. *Lodz Ghetto: Inside a Community under Siege*. New York: Penguin Books.

Ainsztein, Reuben. 1974. *Jewish Resistance in Nazi-Occupied Eastern Europe with a Historical Survey of the Jew as a Fighter and Soldier in the Diaspora*. London: Paul Elek.

Aly, Gotz and Susanne Heim. 2002. *Architects of Annihilation: Auschwitz and the Logic of Destruction*. Trans. by A.G. Blunden. Princeton, NJ: Princeton University Press.

Aly, Gotz, Peter Chroust and Christian Pross. 1994. *Cleansing the Fatherland. Nazi Medicine and Racial Hygiene*. Baltimore: Johns Hopkins University Press.

Apenszlak, Jacob, ed. 1982. *The Black Book of Polish Jewry: An Account of the Martyrdom of Polish Jewry under the Nazi Occupation*. Co-editors Jacob Kenner, Dr. Isaac Lewin, Dr. Moses Polakiewicz. New York: Howard Fertig.

Arad, Yitzhak. 1979. *The Partisan: From the Valley of Death to Mount Zion*. New York: Holocaust Library.

Arad, Yitzhak. 1980. *Ghetto in Flames: The Struggle and Destruction of the Jews in Vilna in the Holocaust*. Jerusalem: Yad vaShem, Martyrs' and Heroes' Remembrance Authority.

Arendt, Hannah. 1977. *Eichmann in Jerusalem*. New York: Penguin.

Armstrong, John A. 1964. *Soviet Partisans in World War II*. Madison: Wisconsin University Press.

Avigdor, Isaac C. 1975. *From Prison to Pulpit: Sermons and Stories*. Hartford, CT: Hanav Publishing.

Bauer, Yehuda. 1973. *They Chose Life: Jewish Resistance in the Holocaust*. New York: American Jewish Committee, Institute of Human Relations.

Bauer, Yehuda. 1979. *The Jewish Emergence from Powerlessness*. Toronto: University of Toronto Press.

Baumel, Judith Tydor. 1998. *Double Jeopardy. Gender and the Holocaust*. Portland, OR: Vallentine Mitchell.

Benshalom, Rafi. 2001. *We Struggled for Life: Zionist Youth Movements in Budapest, 1944*. New York: Gefewn Publishing House.

Berg, Morris. 1945. *Warsaw Ghetto Diary*. New York: L.B. Fisher.

Berger, Ronald J. 1995. *Constructing a Collective Memory of the Holocaust: A Life History of Two Brothers' Survival*. Niwot, CO: University of Colorado Press.

Berkovits, Eliezer. 1973. *Faith after the Holocaust*. New York: Ktav Publishing House, Inc.

Bettelheim, Bruno. 1975. *The Empty Fortress: Infantile Autism and the Birth of the Self*. New York: Free Press.

Bielawski, Shraga Feivel. 1991. *The Last Jew from Wegrow: The Memoirs of a Survivor of the Step-by-Step Genocide in Poland*. Ed. by Louis W. Liebovich. New York: Praeger.
Black Book. August 15, 1951. Newsletter, 'Poland Fights.' See Apenszlak, 1982, above.
Blatt, Thomas Toivi. 1997. *From the Ashes of Sobibor: A Story of Survival*. Foreword by Christopher R. Browning. Evanston, IL: Northwestern University Press.
Blumental, Nachman. 1963. 'Magical Thinking among the Jews during the Nazi Occupation'. In *Yad vaShem Studies on the European Jewish Catastrophe and Resistance*. Eds. Nathan Eck and Arieh Leon Kubovy. Vol. 5.
Blumenthal, David R. 1953. *Facing the Abusing God: A Theology of Protest*. Louisville, KY: Westminster/John Knox Press.
Bokser, Ben Zion. 1981. *The Jewish Mystical Tradition*. New York: The Pilgrim Press.
Bondy, Ruth. 1989. *'Elder of the Jews': Jakob Edelstein of Theresienstadt*. Trans. by Evelyn Abel. New York: Grove Press.
Browning, Christopher. 1992. *Ordinary Men: Reserve Police Battalion 101 and the Final Solution in Poland*. New York: Harper Perennial.
Buber, Martin. 1960. *The Origin and Meaning of Hasidism*. New York: Harper and Row.
Cesarani, David, ed. 1997. *Genocide and Rescue: The Holocaust in Hungary*. New York: Berg.
Cholawski, Shalom. 1980. *Soldiers from the Ghetto*. San Diego: A.S. Barnes & Co., Inc.
Cohen, Asher and Yehoyakim Cochavi, eds. 1995. *Studies on the Shoah: Zionist Youth Movements during the Shoah*. Trans. by Ted Gorelick. New York: Peter Lang.
Cohen, Dov and Jack Kagan. 1998. *Surviving the Holocaust with Russian Jewish Partisans*. Portland, OR: Vallentine Mitchell.
Cohen, Israel. 1992. *Vilna*. Philadelphia: The Jewish Publication Society of America (1943).
Cohen, Rich. 2000. *The Avengers*. New York: A.A. Knopf.
Davis, Mac. 1945. *Jews Fight Too*. New York: Jordon Publishing Co.
Dawidowicz, Lucy S. 1989. *From That Place and Time: A Memoir 1938-1947*. New York: W.W. Norton.
Dean, Martin. 2000. *Collaboration in the Holocaust: Crimes of the Local Police in Belorussia and Ukraine, 1941–44*. New York: St. Martin's Press.
Dobroszycki, Lucjan and Dabrowska, Danuta, eds. 1994. *The Chronicle of the Lodz Ghetto*. Trans. Richard Lourie. New Haven: Yale University Press.
Duffy, Peter. 2003. *The Bielski Brothers: The True Story of Three Men Who Defied the Nazis, Saved 1200 Jews, and Built a Village in the Forest*. New York: HarperCollins.
Eckman, Lester Samuel and Chaim Lazar. 1977. *The Jewish Resistance: The History of the Jewish Partisans in Lithuania and White Russia during the Nazi Occupation, 1940–1945*. New York: Shengold.
Eisen, George. 1988. *Children and Play in the Holocaust: Games among the Shadows*. Amherst: University of Massachusetts Press.

Eisenberg, Azriel Louis. 1982. *The Lost Generation: Children in the Holocaust*. New York: Pilgrim Press.
Eliach, Yaffa. 1982. *Hasidic Tales of the Holocaust*. New York: Oxford University Press.
Elkins, Michael. 1971. *Forged in Fury*. New York: Ballantine.
Fanon, Frantz. 1988. *The Wretched of the Earth*. Trans. by Constance Farrington. New York: Grove-Atlantic.
Ferencz, Benjamin B. 1979. *Less than Slaves: Jewish Forced Labor and the Quest for Compensation*. Cambridge: Harvard University Press.
Fisher, Josey G. 1991. *The Persistence of Youth: Oral Testimonies of the Holocaust*. Westport, CT: Greenwood Press.
Freud, Sigmund. 1975. *The Future of an Illusion*. Trans. and ed. by James Strachey. New York: Norton.
Gar, Joseph. 1966. *Bibliography of Articles on the Catastrophe and Heroism in Yiddish Periodicals*. Joint Documentary Projects, No. 9. New York: Yadvashem and YIVO.
Gelman, Charles. 1989. *Do Not Go Gentle: A Memoir of Jewish Resistance during the Holocaust, 1941–1945*. Hamden, CT: Archon Books.
Gitelman, Zvi, ed. 1997. *Bitter Legacy: Confronting the Holocaust in the USSR*. Bloomington: Indiana University Press.
Glas-Wiener, Sheva. 1983. *Children of the Ghetto*. Trans. by Sheva Glas-Wiener and Shirley Young. Fitzroy, Australia: Globe Press.
Glass, James M. 1997. *'Life Unworthy of Life': Racial Phobia and Mass Murder in Hither's Germany*. New York: Basic Books.
Goldhagen, Daniel Jonah. 1996. *Hitler's Willing Executioners: Ordinary Germans and the Holocaust*. New York: Alfred A. Knopf.
Gottfarstein, Y. 1971. 'Kiddush haShem in the Holocaust Period'. In *Jewish Resistance during the Holocaust*. Proceedings of the Conference on the Manifestations of Jewish Resistance, Jerusalem, April 7–11, 1968. Jerusalem: Yad vaShem.
Gottlieb, Freema. 1989. *The Lamp of God: A Jewish Book of Light*. Northvale, NJ: Jason Aronson, Inc.
Graf, Malvina. 1989. *The Krakow Ghetto and the Plaszow Camp Remembered*. Tallahassee: Florida State University Press.
Granatstein, Yechiel. 1986. *The War of a Jewish Partisan: A Youth Imperiled by His Russian Comrades and Nazi Conquerors*. Trans. by Charles Wengrov. Brooklyn: Mesorah.
Gross, Jan T. 1988. *Revolution from Abroad: The Soviet Conquest of Poland's eastern Ukraine and Western Belorussia*. Princeton, NJ: Princeton University Press.
Gross, Jan Tomasz. 2001. *Neighbors: The Destruction of the Jewish Community in Jedwabne, Poland*. Princeton, NJ: Princeton University Press.
Grossman, Chaika. 1987. *The Underground Army: Fighters of the Bialystok Ghetto*. New York: Holocaust Library.
Grubsztein, Meir and Moshe M. Kohn, eds. 1971. *Jewish Resistance during the Holocaust: Proceedings of the Conference on Manifestations of Jewish Resistance*. Trans. by Varda Esther Bar-On, Moshe M. Kohn, Yehezkel Cohen and Cecil Hyman, April 7–11, 1968. Jerusalem: Yad vaShem.

Gruenwald, Ithamar. 1980. *Apocalyptic and Markavah Mysticism.* Leiden: E.J. Brill.
Gutman, Israel. 1994. *Resistance: The Warsaw Ghetto Uprising.* Boston: Houghton Mifflin.
Gutman, Israel and Michael Berenbaum. 1998. *Anatomy of the Auschwitz Death Camp.* Bloomington: Indiana University Press.
Gutman, Yisrael. 1988. *Fighting among the Ruins: The Story of Jewish Heroism during World War II.* Washington, DC: B'nai B'rith Books.
Haas, Peter J. 1996. *Responsa: Literary History of a Rabbinic Genre.* Atlanta: Scholars Press.
Hilberg, Raul. 1985. *The Destruction of the European Jews*, 2 vols. New York: Holmes and Meier.
Hilberg, Raul. 1995. *Patterns of Jewish Leadership in Nazi Europe, 1933–1945.* New York: Holmes and Meier.
Hill, Benno Muller. 1988. *Marvelous Science: Elimination by Scientific Selection of Jews, Gypsies and Others, Germany 1933–1945.* Trans. by George R. Fraser. New York: Oxford University Press.
Hillesum, Etty. 1987. *Letters from Westerbork.* Trans. by Arnold J. Pomerans. London: Cape.
Hirshler, Gertrude and Shimon Zucker, eds. 1981. *The Unconquerable Spirit: Vignettes of the Religious Spirit the Nazis Could Not Destroy.* Trans. by Gertrude Hirshler. New York: Mensora Publications.
Holliday, Laurel. 1995. *Children in the Holocaust and World War II: Their Secret Diaries.* New York: Pocket Books.
Horn, Joseph. 1996. *Mark it with a Stone.* New York: Barricade Books.
Huberband, Shimon. 1987. *Kiddush haShem: Jewish Religious and Cultural Life in Poland during the Holocaust.* Ed. by Jeffrey S. Gurock and Robert S. Hirt. Trans. by David E. Fishman. Hoboken, NJ: Ktav Publishing House, Inc.
Jacobs, Steven L., ed. 1993. *Studies in the Shoah, Contemporary Jewish Religious Responses to the Shoah,* Vol. V. Lanham, MD: University Press of America.
Johnson, Eric A. 1999. *Nazi Terror: The Gestapo, Jews and Ordinary Germans.* New York: Basic Books.
Kaddushim, Max. 1952. *The Rabbinic Mind.* New York: Jewish Theological Seminary.
Kagan, Jack. 1998. *Surviving the Holocaust with Russian Jewish Partisans.* New York: Vallentine Mitchell.
Kahn, Leon. 1978. *No Time to Mourn: A True Story of a Jewish Partisan Fighter.* Vancouver: Laurelton Press.
Kaplan, Chaim Aron. 1999. *The Scroll of Agony: The Warsaw Diary of Chaim A. Kaplan.* Trans. and ed. by Abraham I. Katsh. Bloomington: Indiana University Press.
Katz, Steven T. 1983. *Post-Holocaust Dialogues: Critical Studies in Modern Jewish Thought.* New York: New York University Press.
Kermish, Joseph, ed. 1986. *To Live with Honor and Die with Honor! Selected Documents from the Warsaw Ghetto Underground Archives. O.S. [Oneg Shabbath].* Jerusalem: Yad vaShem.

Kirschner, Robert, trans. 1985. *Rabbinic Responsa of the Holocaust Era*. New York: Schocken Books.
Klibanski, Bronia. 'The Underground Archives of the Bialystok Ghetto'. In *Yad vaShem. Studies on the European Jewish Catastrophe and Resistance*, Vol. II. Ed. by Shaul Esh. Jerusalem: Ktav Publishing House, Inc.
Klukowski, Zygmunt. 1953. *Diary from Ten Years of Occupation, 1939–1944*. Ed. by Andrew Klukowski and Helen Klukowski May. Trans. by George Klukowski. Chicago: University of Chicago Press.
Kohn, Nahum and Howard Roiter. 1980. *A Voice from the Forest*. New York: Holocaust Library.
Korczak, Janusz. 1978. *The Ghetto Years, 1939–1942*. Trans. by Jerzy Bachrack and Barbara Krzywicka [Vedder]. New York: Holocaust Library.
Kornbluth, William. 1994. *Sentenced to Remember: My Legacy of Life in Pre-1939 Poland and Sixty-Eight Months of Nazi Occupation*. Ed. by Carl Calendar. Bethlehem, PA: Lehigh University Press.
Kosinski, Jerzy. 1976. *The Painted Bird*. New York: Grove Press.
Kowalski, Isaac. 1969. *A Secret Press in Nazi Europe: The Story of a Jewish United Partisan Organization*. New York: Central Guide Publishers.
Kowalski, Isaac, ed. 1984. *Anthology on Armed Jewish Resistance, 1939–1945*. Intro. Yitzhak Arad. Brooklyn: Jewish Combatants Publishers House.
Kraemer, David. 1995. *Responses to Suffering in Classical Rabbinic Literature*. New York: Oxford University Press.
Krakowski, Shmuel. 1984. *The War of the Doomed: Jewish Armed Resistance in Poland, 1942–1944*. Trans. by Orah Blaustein. New York: Holmes and Meier.
Krall, Hanna. 1986. *Shielding the Flame: An Intimate Conversation with Dr. Marek Edelman, the Last Surviving Leader of the Warsaw Uprising*. Trans. by J. Stansinska and C. Weschler. New York: Henry Holt and Company.
Kruk, Herman. 2002. *The Last Days of the Jerusalem of Lithuania: Chronicles from the Vilna Ghetto and the Camps, 1939–1945*. Ed. and intro. by Benjamin Harshaw. Trans. by Barbara Harshaw. New York: YIVO Institute for Jewish Research.
Laing, R.D. 1978. *The Divided Self*. New York: Penguin.
Lambert, Gilles. 1974. *Operation Hazalah: How Young Zionists Rescued Thousands of Hungarian Jews in the Nazi Occupation*. Trans. by Robert Bullen and Rosette Letellier. New York: The Robbs-Merrill Company, Inc.
Landesman, David, trans. 1996. *As the Rabbis Taught: Studies on the Haggadot of the Talmud, A Tishah B'Av Reader*. Northvale, NJ: Jason Aronson, Inc.
Langer, Lawrence L. 1991. *Holocaust Testimonies: The Ruins of Memory*. New Haven, CT: Yale University Press.
Langer, Lawrence. 1995. *Admitting the Holocaust: Collected Essays*. New York: Oxford University Press.
Laska, Vera. 1985. *Nazism, Resistance and Holocaust during World War II: A Bibliography*. Metuchen, NY: Scarecrow Press.
Lazare, Lucien. 1996. *Rescue as Resistance: How Jewish Organizations Fought the Holocaust in France*. Trans. by Jeffrey M. Green. New York: Columbia University Press.

Levi, Primo. 1989. *The Drowned and the Saved.* Trans. by Raymond Rosenthal. New York: Random House.

Levin, Allan. 1998. *Fugitives of the Forest: The Heroic Story of the Jewish Resistance and Survival during the Second World War.* New York: Stoddart.

Levin, Dov. 1985. *Fighting Back: Lithuanian Jewry's Armed Resistance to the Nazis 1941–1945.* Trans. by Moshe Kohn and Dina Cohen. New York: Holmes and Meier.

Levine, Ellen. 2000. *Darkness over Denmark: The Danish Resistance and the Rescue of the Jews.* New York: Holiday House.

Lewin, Abraham. 1988. *A Cup of Tears: A Diary of the Warsaw Ghetto.* Ed. by Antony Polonsky. Trans. by Christopher Hutton. Oxford: Basil Blackwell.

Lifton, Robert Jay. 2000. *The Nazi Doctors: Medical Killing and the Psychology of Genocide.* New York: Basic Books.

Maimonides, Moses. 1982. *Treatise on Resurrection.* Trans. by Fred Kosner. New York: Ktav Publishing House, Inc.

Marrus, Michael, ed. 1989. *Jewish Resistance to the Holocaust: Selected Articles.* Westport, CT: Meckler.

McFarland-Icke, Bronwyn Rebekah. 1999. *Nurses in Nazi Germany: Moral Choice in History.* Princeton, NJ: Princeton University Press.

Mierzejewski, Alfred C. 2001. 'A Public Enterprise in the Service of Mass Murder: the Deutsche Reichsbahn and the Holocaust', *Holocaust and Genocide Studies*, vol. 15, no. 1, Spring.

Mishell, William M. 1988. *Kaddish for Kovno: Life and Death in a Lithuanian Ghetto, 1941–1945.* Chicago: Chicago University Press.

Nirenstein, Albert. 1959. *A Tower from the Enemy: Contributions to a History of Jewish Resistance in Poland.* Trans. from the Polish, Yiddish and Hebrew by David Neiman; from the Italian by Mervyn Savill. New York: Onion Press.

Novitch, Miriam. 1980. *Sobibor, Martyrdom and Revolt: Documents and Testimonies.* New York: Holocaust Library.

Nusan, Jack, ed. 1982. *Jewish Partisans: A Documentary of Jewish Resistance in the Soviet Union during World War II.* Trans. by the Magal Translation Institute. Washington, DC: University of America Press.

Oppenheim, Michael. 1997. *Speaking/Writing of God: Jewish Philosophical Reflections on the Life with Others.* Albany, NY: State University of New York Press.

Orenstein, Henry. 1987. *I Shall Live: Surviving against All Odds, 1939–1945.* Oxford: Oxford University Press.

Oshry, Ephraim. 1995. *The Annihilation of Lithuanian Jewry.* Trans. by Y. Leiman. New York: The Judaica Press, Inc.

Pankiewicz, Tadeusz. 1987. *The Cracow Ghetto Pharmacy.* Trans. by Henry Tilles. New York: Holocaust Library.

Peleg-Marianska, Miriam, and Mordecai Peleg. 1991. *Witnesses: Life in Occupied Krakow.* New York: Routledge.

Polen, Nehemia. 1994. *The Holy Fire: The Teachings of Rabbi Kalonymus Kalman Shapira, the Rebbe of the Warsaw Ghetto.* Northvale, NJ: Jason Aronson, Inc.

Rabinowicz, Tzvi, ed. *The Encyclopedia of Hasidim*. Northvale, NJ: Jason Aronson.
Rakovsky, Puah. 2002. *My Life as a Radical Jewish Woman: Memoirs of a Zionist Feminist in Poland*. Trans. by Barbara Harshav and Paula E. Nyman. Bloomington: Indiana University Press.
Ringelblum, Emmanuel. 1958. *Notes from the Warsaw Ghetto: the Journal of Emmanuel Ringelblum*. Ed. and trans. by Jacob Sloan. New York: McGraw Hill.
Ringelblum, Emmanuel. 1976. *Polish–Jewish Relations during the Second World War*. Ed. by Joseph Kermish and Samuel Krakowski. Trans. by Dafna Allon, Danuta Dbrowska and Diana Keren. New York: Howard Fertig.
Robinson, Jacob and Philip Friedman. 1960. *Guide to Jewish History under Nazi Impact: Joint Documentary Projects*, Bibliographical Series, No. 1. New York: YadVashem and YIVO.
Roland, Charles G. 1992. *Courage under Siege: Starvation, Disease and Death in the Warsaw Ghetto*. New York: Oxford University Press.
Rosenbaum, Irving J. 1976. *Holocaust and Halakhah*. Jerusalem: Ktav Publishing House, Inc.
Rosenberg, Bernhard H. and Fred Heuman, eds. 1992. *Theological and Halakhic Reflections on the Holocaust*. Hoboken, NJ: Ktav Publishing House, Inc.
Rossino, Alexander B. 1997. 'Destructive Impulses: German Soldiers and the Conquest of Poland'. *Holocaust and Genocide Studies*, Vol. VII, No. 3 (Winter).
Rotem, Simha. 1994. *Memoirs of a Warsaw Ghetto Fighter: The Past within Me*. Trans. from the Hebrew and ed. by Barbara Harshav. New Haven, CT: Yale University Press.
Rousseau, Jean-Jacques. 1951. *The Social Contract and Discourses*. Trans. by G.D.H. Cole. New York: E.P. Dutton.
Rudashevski, Yitzhok. 1973. *The Diary of the Vilna Ghetto: June 1941–April 1943*. Trans. by Percy Matenko. Jerusalem: Ghetto Fighters' House.
Rudavsky, Joseph. 1987. *To Live with Hope: To Die with Dignity*. Lanham, MD: University Press of America.
Scharf, Rafael F., ed. 1993. *In the Warsaw Ghetto: Summer 1941. Photographs by Willy Georg*. New York: Aperture Foundation.
Schindler, Pesach. 1990. *Hasidic Responses to the Holocaust in the Light of Hasidic Thought*. Hoboken, NJ: Ktav Publishing House, Inc.
Schoenfeld, Joachim. 1985. *Holocaust Memoirs: Jews in the Lwow Ghetto, the Janowski Concentration Camp, and as Deportees in Siberia*. Hoboken, NJ: Ktav Publishing House, Inc.
Scholem, Gershon G. 1954. *Major Trends in Jewish Mysticism*. New York: Schocken Books.
Schuchter-Shalomi, Zalman. 2003. *Wrapped in a Holy Flame: Teachings and Tales of the Hasidic Masters*. San Francisco: Jossey-Bass.
Schulman, Faye. 1995. *A Partisan's Memoir: Woman of the Holocaust*. Toronto: Second Story Press.
Schwarzbart, Isaac I. 1953. *The Story of the Warsaw Ghetto Uprising: Its Meaning and Message*. New York: World Jewish Congress, Organization Department.

Shapira, Rabbi Kalonymus Kalman. 1991. *A Student's Obligation: Advice from the Rebbe of the Warsaw Ghetto.* Trans. by Michael Odenheimer. Northvale, NJ: Jason Aronson, Inc.
Smolar, Hersh. 1989. *The Minsk Ghetto: Soviet-Jewish Partisans against the Nazis.* New York: Holocaust Library.
Stoltzfus, Nathan. 1996. *Resistance of the Heart: Intermarriage and the Rosenstrasse Protest in Nazi Germany.* New York: W.W. Norton.
Suhl, Yuri, ed. and trans. 1975. *They Fought Back: The Story of Jewish Resistance in Nazi Europe.* New York: Schocken Books.
Sutin, Jack and Rochelle Sutin. 1995. *Jack and Rochelle: A Holocaust Story of Love and Resistance.* St. Paul, MN: Graywolf Press.
Szner, Zoe, ed. 1958. *Extermination and Resistance: Historical Records and Source Material.* Haifa: Ghetto Fighters House.
Szwajger, Adina Blady. 1990. *I Remember Nothing More: The Warsaw Children's Hospital and the Resistance.* Trans. by Tasja Darowska and Danusia Stok. New York: Pantheon.
Tec, Nechama. 1993. *Defiance: the Bielski Partisans.* New York: Oxford University Press.
Tenenbaum, Joseph. 1952. *Underground: The Story of a People.* New York: Philosophical Library.
Tory, Avraham. 1990. *Surviving the Holocaust: The Kovno Ghetto Diary.* Ed. by Martin Gilbert. Trans. by Jerzy Michalowicz. Cambridge, MA: Harvard University Press.
Trunk, Isaiah. 1977. *Judenrat: The Jewish Councils in Eastern Europe under Nazi Occupation.* New York: Stein and Day.
Trunk, Isaiah. 1979. *Jewish Responses to Nazi Persecution: Collection on Individual Behavior in Extremis.* New York: Stein and Day.
Twersky, Isadore, ed. 1972. *A Maimonides Reader.* New York: Behrman House, Inc.
Unger, Menashe. 1967. *Sefer Kedoshim: Rebeim of Kiddush Hashem* [the Book of Martyrs: Hasidic Rabbis as Martyrs During the Holocaust]. New York: Shul Singer.
Waller, James. 2002. *Becoming Evil: How Ordinary People Commit Genocide and Mass Killing.* New York: Oxford University Press.
Wells, Leon Weliczker. 1995. *Shattered Faith.* Lexington: The University Press of Kentucky.
Werner, Harold. 1992. *Fighting Back: A Memoir of Jewish Resistance in World War II.* Ed. by Mark Werner. New York: Columbia University Press.
Winnicott, Donald W. 1958. *Collected Papers.* London: Tavistock.
Wygoda, Hermann, ed. 1998. *In the Shadow of the Swastika.* Urbana, IL: University of Illinois Press.
Yehude Yaar [Forest Jews]. 1946. *Narratives of Jewish Partisans of White Russia, Tuvia and Zus Bielski and Abraham Viner.* As told to Ben Dor. Tel Aviv: Am Oved.
Yoram, Shalom. 1996. *The Defiant: A True Story.* Trans. by Varda Yoram. New York: St. Martin's Press.
Zable, Arnold. 1991. *Jewels and Ashes.* New York: Harcourt, Brace and Co.

Zimmels, H.S. 1977. *The Echo of the Nazi Holocaust in Rabbinical Literature.* London: Ktav Publishing House, Inc.
Zuckerman, Yitzhak. 1993. *A Surplus of Memory: Chronicles of the Warsaw Ghetto Uprising.* Trans. and ed. by Barbara Harshav. Berkeley: University of California Press.
Zuker, Shimon. 1981. *The Unconquerable Spirit: Vignettes of the Jewish Religious Spirit the Nazis Could Not Destroy.* Trans. by Gertrude Hirschler. New York: Mesorah Publications.

Index

abortions, in partisan camps 84
Achdut Haavoda underground 42
agency, drained 149
Akiba, Zionist underground
  organization 42
Aly, Gotz 118
anger, in ghettos 45
Anielewicz, Mordechai 31
anti-Semitism
  in partisan groups 96–7
  Polish 53–4, 96–7, 98
  Russian 9, 10
  Warsaw 37
Arad, Yitzhak 49, 50, 97
Arendt, Hannah
  criticism of *Judenrate* 19–20
  Eichmann in Jerusalem 20, 161
Auschwitz
  knowledge of 45
  memorial 135
  selections in 142–3
  transportation to 20–2
  underground community of
    resistance 3

Baranowicze region 58
Bauer, Yehuda 93
Begin, Menachem 42
Beitar, Zionist-revisionist
  underground organization 42
belief, power of 15
Bell, Aaron 166
Berger, Schlomo 3, 6
Berkovits, Eliezer 162
Bialystok ghetto 22, 24, 37
  Judenrat in 34
  underground in 45
Biebow, Hans 185*n*
Bielski Brigade 9–13, 55–62, 80, 99
  fighting brigade 60–1

internal conflicts 68–9, 76
interviews with survivors 62–78
relations with Russian
  commanders 58, 59, 89
relations with Soviets and local
  population 62
social hierarchy 56–7, 60–1
use of violence 87
Bielski, Asael 55
Bielski (Bell), Aaron 61, 73–8
Bielski, Sonia 5, 61
  interview with 62–7
Bielski, Tuvia 9–10, 13, 55
  leadership of 56, 59–60, 99–100
Bielski, Zush (Zeisal) 9, 10–13, 55,
  57
  and value of violence 81
Bielski, Zvi 9, 10–11
Blaichman, Frank 14–15
Bledzow, Charles 68
Bleichman, Frank 6
Borowski, Chiena 49
bribes 45
Browning, Christopher 118
Budapest 21, 27
Bund, Jewish labor union 42
Byelorussia
  forests of 37, 55
  Jewish resistance in 93
  see also Bielski Brigade

catatonia, in ghettos 139
Césaire, Aimé, *Les Armes*
  Miraculeuses 86
children 17
  in ghetto hospitals 28–30
  in ghettos 37–8, 40–1
  and *Kiddush haShem* 162
  and *Kinder* selection in Lodz 138
  in partisan camps 84

children – *continued*
  study of Torah 35
  theological status of 132
Chwojnik, Abrasha 49
collaborators 122, 178*n*
  executed by partisans 14, 75
  and partisans 73–4
community
  among partisans 72, 84, 89
  Rousseauian 100, 113, 180*n*
  and self 99–101, 113
  within ghettos 21
compassion 7, 166–7
contingency, role in survival 16
cultural practices 6, 153
  and conditioning 13
  importance to partisans 12

death
  acceptance of (with faith) 105
  envied 145
  and resurrection 121
  see also mortality rates
death camps, knowledge of 34, 45, 131
despair, and spiritual resistance 33–4
diaries 103
Diaspora, tradition of non-violence 52
dislocation 124, 131
dissociation 44, 116–18, 139
  and indifference 163
  as mode of survival 145
D'ror, Zionist youth organization 42
Duffy, Peter, *The Bielski Brothers* 55–6, 62

Edelman, Marek 133
Efrati, Rabbi Shimon 130
Eichmann, Adolf 27
ethics
  dilemmas 124, 133, 155–7
  and survival 68
evil, Manichean 87

faith 6, 7, 33, 119
  and despair 135–6
  in God 105, 129–30
  in Jewish community identity 105
  loss of 139–40, 165–6
  and mysticism 160–3
  power of 151
  and soul-death 137–40
  theological 132–3, 136–7
family, centrality of 21, 24, 94
Fanon, Frantz
  The Wretched of the Earth 81
  theory of violence 85–6, 88–9
Farfel, Siomka 80
fear, control of 75
Fisch, Rabbi Yehezkiah 159, 161
food rationing and supplies
  in ghettos 38, 51–2, 126–7, 185*n*
  malnutrition 38, 175*n*, 185*n*
  soup kitchens 51
forced labor *see* labor brigades
Freud, Sigmund, *The Future of an Illusion* 100–1
Frucht, Major Isidor 49
funerals, in ghettos 123

Gandhi, Mahatma, passive resistance 152
Gaststeiger, Sergeant 38
genocide, principles of 134–5
Gens, Joseph 48, 95
German policy
  as contradiction of economic self-interest 133–4
  ghettoization 13
  as group madness 167–8
  towards ghettos 20–4
  underground understanding of 45
  vocabulary of killing 45–6
Germans
  biological phobia of Jewish bodies 117–19, 125–6, 134, 168
  reaction to resistance 82

ghettoization 13
  effect on inhabitants 37–42, 83–4
ghettos
  administration in 48
  catatonia in 139
  community support within 21
  conditions in 35–6, 37–42, 141, 183n
  deaths in 93, 123, 137–8, 175n, 183n
  demoralization and breakdown in 19–24, 32–4
  despair in 135–6
  diaries of 103
  difficulty of escape from 15, 30, 48, 76, 94
  disintegration in 43, 72–3, 182n
  hospitals in 28–30, 52
  passivity in 83–4
  physical assaults 124–5
  practices of domination in 131
  survival in 153
  and transfer to death camps 131
  see also Judenrate; underground resistance; Vilna; Warsaw
gift-giving 45
Glazman, Joseph 49
Goldhagen, Daniel 118, 133
Gordonia, Zionist youth group 42
greed 68
Gross, Jan, *Neighbors* 37
Grossman, Chaika 34
guilt
  and conscience 130–1
  lack of 3

Halakhah 157
Hanoar Hatzioni, Zionist youth group 43
Hashomer Hatzair, Zionist youth group 43
health
  in partisan groups 61, 79
  practices in ghettos 125–7
Hechalutz, Zionist pioneer youth movement 42

Hekhalot Rabbati 157–8
Hilberg, Raul 19
Hill, Benno Muller 118
Himmler, Heinrich 185n
Hitachdut, Zionist youth group 43
Horn, Joseph 163, 164, 165–6
humanity, disintegration of 144–5
Hungary, extermination of Jews in 21, 27

ideology, role of 44–5
infants
  allowed to die 18, 172n
  killed at birth 24
informers, Jews as 83
Irgun Tzvai Leumi, Zionist-revisionist military wing 42
Israel, state of 74

Jabotinski, Vladimir 42
Jewish Fighting Organization (JFO) 31
Jewish identity 2–3
  among survivors 146–7
  and group identity 113–14
  importance to partisans 11, 33–4, 77
  and rescue of self 147
Jewish Military Association 32
Jewish police 104
  Warsaw 41–2
Jewish Self-Help organization 51
Judenrate (in ghettos)
  collaboration with Germans 19–21, 49–50, 54
  escape discouraged by 15–16, 76
  failure of leadership 84–5
  fear of reprisals 15–16, 30–1, 46–7, 49
  knowledge of death camps 34
  relations with rabbis 108, 155
  relations with underground resistance 31, 53, 85
  view of violence 83–4

Kamm, Ben 6
Kaplan, Chaim A. 182n
Kiddush haShem (martyrdom)
  105–6, 148, 150, 152–4
  children and 162
  and will to live 159
killing
  Bielski Unit's view of 12–13
  morality of 45
Kinderaktion, in ghettos 41, 138
Klingberg, Rabbi Shem 159
Kohn, Nahum 97–8
Korczak, Januz 127–8
Kovner, Abba 31, 49, 50
Kovno ghetto 22, 24, 32–3, 35, 37
  Judenrat 156
  Kinderaktion in 41
  underground 45
Kowalski, Zygmunt 36
Krakow 125–6, 141–2
  role of rabbis in 151
Kroll, Dr 125, 126

labor brigades 131–2, 133–4, 161
Langer, Lawrence 135–6, 142
  survivor testimony 143, 144–9
leadership, covert in ghettos 84–5
Lerman, Miles 5, 19, 23, 24, 107
Levi, Primo 134–5
Levin, Don 93
Lewin, Abraham 38, 42
life, struggle for 39
Lithuania
  forests of 37
  Jewish resistance in 93, 178–9n
  sympathy for Jews' plight 36
  see also Vilna
local populations 178n
  dangers of hiding Jews 30
  inability to trust 36
  leaflets distributed to 31
Lodz ghetto 20, 22, 38, 90, 134, 146, 185n
  Kinder selection 138
  and removal to Auschwitz 20, 185n
  suicides 157, 183–4n

London News Chronicle 52
Lublin 158
luck, role in survival 3, 143, 149
Lvov ghetto 22

madness
  German policy as 167–8
  rabbinical defenses against 103–4
  and violence 89–90
Maidanek, memorial 135, 164
Maimonides, Moses 121–2, 153–4
malbush, role of 57
malnutrition, in ghettos 38, 175n, 185n
Manicheanism 87
Markov Brigade partisans 16
martyrdom see Kiddush haShem
Masada, tradition of 44
mass reprisals
  German policy of 13, 46–7
  Judenrate fear of 15–16, 30–1
Meisels, Rabbi 133, 157–8
memory 147–8
Mendel, Rabbi 33
Mir ghetto 37
  underground 45
morality 149
  corrupted in ghettos 41–2
  of killing 45
  redefined 68
  and resistance 24–5, 44
mortality rates
  among partisans 55, 79, 93
  children 52, 172n
  in ghettos 93, 123, 137–8, 175n
  in labor groups 131–2
  Warsaw 21, 52
muteness 129, 144, 145
mystical theology 158, 159–60
mysticism, and faith 160–3

Nalibocka forest, Byelorussia 57–8
natality, importance of 67–8, 69–70, 146
Nesvizh, Poland, ghetto 27–8

New York Times, report on Vilna
  ghetto 46
Ninth Fort (outside Kovno) 45
Nissenbaum, Rabbi Yitzhak 154
Novogrudek, massacres at 56, 71

Operation Erntefest (Operation
  Harvest Festival) 133
oppression
  and faith 139
  kills self 87–8
Oshmann, Sonya 6
Oshry, Rabbi, Kovno ghetto 35,
  41, 124, 161
Ozer, Rabbi Chaim 152–3

partisans
  command organization 79
  community support among 72,
    84, 89
  demographics of 93
  lack of guilt 3–4, 5
  and morality of killing 45
  political organization 83, 100
  recruitment of fighters 24
  revenge and retaliation 6–7,
    81–2, 99
  size of units 80
  Soviet (Russian) groups 9, 10,
    79, 91–2, 95–6
  survival in 18–19
  use of violence 7, 83–4
  see also Bielski Brigade
Platon, General 58, 59
Plazow labor camp 141, 147
Poalei Zion, Zionist labor section
  42
Poland
  Hasidic Judaism in 151
  Jewish resistance in 93, 180*n*
  occupation of 38–9, 122
  see also Lodz; Warsaw
Poles
  anti-Semitism 53–4, 96–7, 98
  brutality towards Jews 36, 92–3
  proscribed from trading with Jews
    51

  segregation from Jews 125–6
  view of Jews' plight 36, 53–4
Polewka, Adam 53
Polish Army (AK), hatred of Jews
    97
Polish resistance, Jewish contacts
  with 52–3
political organization, as moral
  authority 44
political resistance 30, 31–2
  in ghetto undergrounds 42–3
  and orthodox theology 151
  spiritual resistance as 105
Ponary (execution site near Vilna)
  34, 39
  knowledge of 45, 47
prayer 112–13, 150

rabbinical response 103–4
  see also spiritual resistance
rabbinical teaching 7
  responsa 155–7, 162
rabbis
  courage of 33, 105–6
  and ethical dilemmas 124, 133,
    182*n*
  in Krakow 151
  moral authority of 108–10
  and occupation of Poland 122
  protection of sacred objects
    106
  relations with *Judenrate* 108
  and resistance 104, 154
  see also Oshry; Shapira
Rasheen, Vernon 6
reason, and faith 137
refugees 38–9
  in Warsaw 52
Reichsbahn, role of 177*n*
religion
  and self 100–1, 136
  see also theology
religious identity 107–8
religious practices 153
  in death camps 154, 182*n*
  in forests 144
  in ghettos 35, 39–40, 107, 122

religious practices – *continued*
  importance of, to survivors
    69–70, 146
  as response to despair 152
  remnant mentality, Warsaw 53
  rescue
    imagery of 44, 45
    as motivation of resistance
      fighters 55, 59, 99–100
  resistance
    passive 152
    revision of morality 24–5, 44
    to group madness 167
    see also Bielski Brigade; partisans;
      underground resistance;
      Warsaw Uprising
Resnik, Nisr 49
resurrection of the dead 121
retaliation 6
revenge 6, 44, 55, 71–2
  personal commitment to
    14–15
  and violence 81–2, 99
Ringelblum, Emmanuel 32, 52–3,
  104, 148, 151, 162
Rudashevski, Yitzhok 39
Rumkowski, Chaim 20

sabotage 50, 82, 92
sacrifice, religious 160–1
Sanctification of the Name of the
  Lord 105, 150
Schindler, Oscar 141–2
Schulman, Faye 81
self
  and community 99–101
  and dissociation 163
  and faith in God 128–31
  and group action 79, 82
  Hasidic 160–1
  killed by oppression 87–8
  liberation of 25
  preservation of 104, 110–14
  reclaimed by violence 84
  resistor 21
  and silence 129, 144
  spiritual resistance and 147–8

and survival 145–6
transformation of 143, 146–7
Shapira, Rabbi Kalonymus Kalman
  5, 6, 121, 149, 184*n*
  consolation of faith 135, 136,
    139–40, 150
  and sacrifice 132
  theological writings 114–16,
    118–19, 128–30, 134
Shapiro, Rabbi Abraham 156
shoes, children's 135, 164
Sierpe, Poland 123
Sobibor 21, 132
  escape attempt 149
soul-death, and faith 137–40
Soviet Union
  Jews in 11
  and partisan brigades 58–9
  see also Byelorussia
Speer, Albert 185*n*
spiritual resistance 7, 32–3
  attitude of Bielskis to 10–11, 12
  and facing death 181*n*
  and faith in Jewish identity 105
  as psychological refuge 104–5,
    110–14, 163–7
  rejected by partisans 15, 105,
    107
  and rescue of self 147–8
  Shapira's writings 114–16,
    118–19
Strehn, Corporal Mathias 38
study 35, 152–3
  as religious act 166
suffering
  idealized 163
  righteousness of 164–5
suicides, Lodz ghetto 157,
  183–4*n*
survival
  among partisans 100
  and ethics 68
  and faith 143–4
  and luck 3, 143, 149
  role of contingency 16, 153–4
  spiritual resistance and 149–50
  and will to live 159

survivors
 and guilt 141–2
 pride in children 4–5
 stories 1–2, 103
 see also partisans
Switzerland 39
Szwajger, Dr Adina Blady 28, 29, 40

Tannenbaum, Mordechai 31
Tec, Nechama, *Defiance: The Bielski Partisans* 55–6, 66
theology 33, 128
 concept of divine justice 166–7
 and concept of fate 105–6
 Hasidic Judaism 110, 151, 152, 160–1
 mystical 158, 159–60
 role of 150, 153
 status of infants and children in 132
Theresienstadt ghetto 22
Torah, study of 35, 152–3
Tory, Abraham 45
Trakinski, Simon 11, 15–18
transport, and dislocation 124, 131, 177*n*
Treblinka 21, 34, 45, 132

underground resistance 3
 charismatic leadership 84–5
 in ghettos 22–3, 24, 30–1, 34–7, 42–8
 interdependence in 43–4
 links with partisans 50
 and morality of killing 45
 in Vilna 45–6, 47, 48–51
 Zionist units 27
United Partisan Organization 47, 50–1
United States, survivors' lives in 13, 18

vengeance *see* revenge
Vilna ghetto 16, 22, 24, 37
 Aktion (1943) 95–6
 conditions in 39–40
 escape to forests from 95–8
 Judenrat policy 46–7, 48–9, 50, 95
 reprisals 93–4
 underground organization 45–6, 47, 48–51, 95
Viner, Abraham 60
violence
 discriminatory use of, by partisans 83–4
 and madness 89–90
 psychological value of 81–2, 84
 reciprocal 86
 and recovery of self 85–94, 100–1

Wallenberg, Raul 82
Warsaw
 deaths (1942) 21, 157, 172*n*
 deportations (1942) 53, 158–9
 ghetto 22, 24, 32, 38, 127–8, 141
 Jewish police 41–2
 refugees in 52
 soup kitchens 51
 underground resistance 31–2, 42–3
Warsaw Uprising (1943) 23–4, 45
weapons
 acquisition of 49, 53
 required by partisan groups 95–6, 98
Westerbork (Netherlands) 22
will
 loss of 144, 165–6
 to live 159
Wittenberg, Itzik 46, 48
women, in Bielski Brigade 59, 65
work brigades 131–2, 133–4, 161
work permits 39, 94
workshops, at Bielski base area 57–8
Wylezynska, Aurealia 53

Yellin, Chaim 31, 33
Yitzhak, Rabbi Levi 159–60

Zaczepice village, Byelorussia 83
Zemba, Rabbi Menachem 159, 162
Zionist-revisionists 42
Zionists, underground units 27, 30, 31–2, 42–3
Zukerman, Yitzhak 31

The manufacturer's authorised representative in the EU is Springer Nature Customer Service Centre GmbH, Europaplatz 3, 69115 Heidelberg, Germany. If you have any concerns regarding our products, please contact ProductSafety@springernature.com

Printed and bound by CPI Group (UK) Ltd, Croydon, CR0 4YY

26/03/2026

02078853-0010